T0345280

Power System Transients

In this textbook, a variety of transient cases that have occurred or are possible to occur in power systems are discussed and analyzed. It starts by categorizing transients' phenomena and specifying unfavorable situations in power systems raised by transients. It then moves on to different protective measures that have been implemented in the system to prevent disasters caused by those transients. It also explains different methodologies used to analyze transients in power systems.

This book discusses the modeling of components very extensively and provides analysis cases to assess a wide variety of transients, their possible effects on the system, and the types of protection commonly used for each case, along with methods for designing a sound protection system.

FEATURES

- Detailed models of system components along with power systems computer-aided design (PSCAD) implementation and analysis
- Comprehensive reference of transient cases in power systems along with design considerations and protective solutions
- The cases are not limited to classical transients such as lightning strikes and switching, but rather the book discusses transient cases that power system operators and engineers have to deal with, such as ferroresonance, in detail, accompanied by computer simulations
- A chapter on original materials related to transformer windings with induced traveling waves

Power System Transients: Modelling Simulation and Applications provides a comprehensive resource to mainly educate graduate students in the area of power system transients. It also serves as a reference for industry engineers challenged by transient problems in the system.

Power System Transients

Modelling Simulation and Applications

Gevork B. Gharehpetian

Atousa Yazdani

Behrooz Zaker

CRC Press
Taylor & Francis Group
Boca Raton London New York

CRC Press is an imprint of the
Taylor & Francis Group, an **informa** business

MATLAB® is a trademark of The MathWorks, Inc. and is used with permission. The MathWorks does not warrant the accuracy of the text or exercises in this book. This book's use or discussion of MATLAB® software or related products does not constitute endorsement or sponsorship by The MathWorks of a particular pedagogical approach or particular use of the MATLAB® software.

First edition published 2023
by CRC Press
6000 Broken Sound Parkway NW, Suite 300, Boca Raton, FL 33487-2742

and by CRC Press
Park Square, Milton Park, Abingdon, Oxon, OX14 4RN

CRC Press is an imprint of Taylor & Francis Group, LLC

© 2023 Gevork B. Gharehpetian, Atousa Yazdani and Behrooz Zaker

ISBN: 978-1-032-18558-3 (hbk)
ISBN: 978-1-032-18559-0 (pbk)
ISBN: 978-1-003-25513-0 (ebk)

DOI: 10.1201/9781003255130

Typeset in Times
by SPi Technologies India Pvt Ltd (Straive)

Contents

Preface

Over the years, utility engineers have been challenged by dealing with many of the components discussed in the book.

Examples of those cases are not limited to line energization, equipment switching, and many more contained in the book. Therefore, this will be a great reference for their analysis and design adjustments. Also, the book serves the educational needs of graduate students in the field of power engineering. The author has taught the course for almost 20 years and now has received praise from European industries involved in the concept. Throughout these years, the students have been challenged by finding a comprehensive book that covers a large number of concepts in the area. The book provides a concrete reference for people who are working in modeling and design of power systems considering transients. The mentioned group of people can be as large as graduate students, consultants, utility engineers, and more.

The book provides a comprehensive resource to mainly educate graduate students in the area of power system transients. It also serves as a reference for industry engineers challenged by transient problems in the system. The book starts by categorizing transients' phenomena and specifying unfavorable situations in power systems raised by transients. Then, it moves on to different protective measures that have been implemented in the system to prevent disasters caused by those transients. It also explains different methodologies used to analyze transients in power systems. A variety of transient cases that occurred or are possible to occur in power systems are discussed and analyzed in this textbook. The book discusses the modeling of components very extensively and provides analysis cases to assess a wide variety of transients, their possible effects on the system, and types of protections commonly used for each case along with methods in designing a sound protection system.

Modeling of system elements such as transmission lines and cables with frequency dependency consideration is an integral part of the book. Detailed PSCAD models of the system components are presented and are used in system-level analyses. Modeling will be done in a multi-phase platform accommodating mathematical models of various elements such as overhead lines, cables, transformers, and so on. These models are built to accommodate system transients in time scales as small as microseconds. Various issues caused by transients in power grids are explored and simulated in detail such as transients in connection nodes of cables and transmission lines, ferroresonance, and several other examples like this.

Low-frequency and mid-frequency transients such as inrush current, ferroresonance, faulted line transients, and magnetizing current chopping are discussed and analyzed. Also, high-frequency cases such as lightning strikes and switching transients are discussed, modeled, and analyzed.

The book covers deterministic and stochastic simulations and calculation methodologies and discusses how they are different, but eventually, the goal of presenting the book is to use deterministic approaches in studying a variety of transient cases in power systems.

The abovementioned subjects have been organized and presented in the following eight chapters:

Chapter 1 provides an overview of different transient conditions and categorizes them based on the duration of time they affect the system and also the range of effects (i.e., the amount of voltage rise in the system). It provides details about high voltage and high current transients along with abnormal waveforms and electromechanical transients. This chapter discusses different available solution techniques mentioned above as stochastic and deterministic analysis. Here, the authors elaborate on the differences between those methodologies and discuss appropriate places for each type of analysis. Also, the chapter discusses some of the prevalent shortcomings related to the modeling platforms and proposes possible solutions.

Chapter 2 discusses the traveling wave theory. Different line models are considered in this chapter such as lossless and lossy components. Further on, details required for analyzing overvoltages considering the traveling wave concept are organized and explained as reflection rules for sinusoidal waves and step waves.

Chapter 3 discusses the lattice diagram. Lattice diagram is an old and simplified way to analyze the effects of transients (i.e., overvoltages on the system) considering the concept of traveling waves. This simplified methodology works best where there is an absence of modern tools for analyzing transients. The authors of this proposal believe that the lattice diagram is a very good method to present the principal concepts of traveling waves to students without using any software. Also, the simplicity lies in the methodology guiding the reader to do a hands-on practice on the matter. It helps the reader understand the rise of voltages in the system due to transients and wave reflections.

Chapter 4 discusses lightning-induced transients and their effects on power systems. In the case of lightning strikes, we consider a thorough approach in discussing stochastic modeling for this specific very fast transient case. In this chapter, we provide several PSCAD cases required for educational purposes on the concept. The result of overvoltages reported by the simulation software will be compared with the results obtained using lattice diagrams as practice points.

Chapter 5 discusses components energization (e.g., sources, lines, and circuit components) and their effects on the system. The stochastic behavior of overvoltages due to line energization and asynchronous closing of three-phase contactors have also been discussed. Each subsection contains a thorough explanation of the case along with multi-phase PSCAD simulations and analyses.

Chapter 6 discusses the de-energization of components (e.g., sources, lines, and circuit components) and their effects on the system. This chapter elaborates on a range of useful definitions required for analyzing the system effects in case of component de-energizations such as Transient Recovery Voltage (TRV) and Rate of Rise of Recovery Voltage (RRRV). In this chapter, we discuss a methodology that is traditionally being used, which is called the Current Injection Method. This method is used as a way to analyze the effects the system has to tolerate in case of component de-energization. Multi-phase PSCAD simulation and analysis cases are presented in the following cases: Fault interruption in a short line, magnetizing current chopping, capacitive current interruption, disconnectors opening, and single-line to ground fault opening. In several cases, hands-on practices have been included comparing the results from the simulation platform to the results obtained by current injecting methodology.

Chapter 7 discusses the transient phenomena inside an important device of a power system, that is, transformers. This chapter can attract design engineers in transformer-manufacturing firms. Here, the focus is on transformers and how they are affected by transient waves. The modeling of transformer windings for transient studies is presented and the connection between insulation design of HV windings and overvoltages due to transients is explained. The model can be used for reactors and electrical machines windings, that is, the concept can be applied to other devices as well. These are important concepts for graduate students and power system device designers.

Chapter 8 discusses a phenomenon in a power system that can cause unfavorable rises in voltage levels. Usually, this phenomenon, called ferroresonance, is typically initiated by a combination of a saturable magnetizing inductance of a transformer and a cable or transmission line with intrinsic capacitors that are connected to the transformer. In most practical situations, ferroresonance results in dominated currents, but in some operating modes, it may cause significantly high values of distorted winding voltage waveform, which is typically referred to as ferroresonance. Although the occurrence of a resonance involves a capacitance and a saturable inductance with a known resonance frequency, there is no definite resonance frequency in this case. In this phenomenon, more than one response is possible for the same set of parameters, and drifts or transients may cause the response to jump from one steady-state response to another. In this chapter, different practical conditions resulting in this phenomenon have been analyzed and discussed.

We summarize the book's outstanding features as follows:

1. In-depth system component modeling and analysis.
2. Utilization of hands-on software tools to provide experimental platforms and enhance learning outcomes.
3. The variety of the concepts explained and examined in the book.

MATLAB® is a registered trademark of The Math Works, Inc. For product information, please contact:

The Math Works, Inc.
3 Apple Hill Drive
Natick, MA 01760-2098
Tel: 508-647-7000
Fax: 508-647-7001
E-mail: info@mathworks.com
Web: http://www.mathworks.com

Authors

Gevork B. Gharehpetian received his PhD degree in Electrical Engineering in 1996 from Tehran University, Tehran, Iran, graduating with First Class Honors. As a PhD student, he received a scholarship from DAAD (German Academic Exchange Service) from 1993 to 1996 and he was with the High Voltage Institute of RWTH Aachen, Aachen, Germany.

He held the assistant professor position at AUT from 1997 to 2003, the position of associate professor from 2004 to 2007, and has been a professor since 2007.

He was selected by the MSRT (Ministry of Science Research and Technology) as the distinguished professor of Iran, by IAEEE (Iranian Association of Electrical and Electronics Engineers) as the distinguished researcher of Iran, by the Iran Energy Association (IEA) as the best researcher of Iran in the field of energy, by the MSRT as the distinguished researcher of Iran, by the Academy of Science of the Islamic Republic of Iran as the distinguished professor of electrical engineering, by National Elites Foundation as the laureates of Alameh Tabatabaei Award and was awarded the National Prize in 2008, 2010, 2018, 2018, 2019, and 2019, respectively. Based on the Web of Science database (2005–2019), he is among the world's top 1% elite scientists according to the ESI (Essential Science Indicators) ranking system.

He is the author of more than 1200 journal and conference papers. His teaching and research interests include smart grids, microgrids, FACTS and HVDC systems, and monitoring of power transformers and their transients.

Atousa Yazdani is an associate professor at California State University, Sacramento. She received her PhD from the Missouri University of Science and Technology in 2009. She received her MSc and BSc in Tehran Polytechnics (2001) and Tehran University (1997), respectively. She has been involved with several consulting projects both in the USA and abroad. She has been instrumental in several utility-scaled projects such as the integration of renewable energy and the implementation of stability algorithms utilizing synchrophasors data. She has many years of industry experience in design, modeling, and analysis of power systems. Her main research interest is in the application of power electronic apparatus in power systems, dynamic system analysis, and testing of power systems with the inclusion of novel technologies.

Behrooz Zaker is an assistant professor at Shiraz University. He received his BSc and MSc from Shiraz University (2011) and Amirkabir University of Technology (2013), respectively. He received his PhD degree in Electrical Engineering in 2018 from Amirkabir University of Technology Tehran, Iran, graduating with First Class Honors. He has many years of industry experience in modeling and parameter estimation of thermal and hydro power plants. His research interests include system identification, power system dynamics, distributed generation systems, and microgrids.

1 Overview

1.1 INTRODUCTION

Power systems, as we know them today, have been implemented to accommodate the electric customers in a safe and reliable way by transmitting the produced energy in power plants through transmission and distribution systems. Generators are devices used to convert mechanical energy into electrical energy; power transmission is performed through copper and aluminum lines in which they carry electric current, and transformers are used to change voltage levels required for transmission and distribution. In such systems, the physical dimensions of elements are much smaller than current and voltage wavelengths, for example, the wavelength is more than 3000 miles for 60 Hz. With this consideration, it is not possible to use lumped models for the electrical system components, which also apply nodal and mesh analysis in a connected group of components (i.e., Kirchhoff's laws for analyzing linear circuits). In AC systems with constant operational frequency of 50 or 60 Hz, phasor analysis is chosen over solving differential equations of the systems. If a number of observers watch system variables and those variables stay constant or unchanged, then the system is operating in a steady-state condition. In the case of power systems, observer is a variable that rotates with the speed of ω, which is the angular frequency of the system. Therefore, such an observer will observe other variables that are rotating with the same speed as a constant.

On the contrary, transients are the conditions in which we cannot find any observers who have the same condition as the steady-state observer. In such conditions, the system is in a transient mode. Transients in power systems include high frequencies up to megahertz. In such conditions, system variables alter very rapidly. Therefore, phasor analysis will render wrong results. Differential equations, which normally define the lumped system, are not useable in this case either. For instance, in a steady-state condition at 50 or 60 Hz frequency, the transformation ratio of a power transformer is the ratio of the number of turns on the secondary winding to the number of turns on the primary winding. But this is not the case when dealing with lightning-induced voltages. In such conditions, stray capacitors between the primary and secondary windings will influence the transformation of voltage between the primary and the secondary coils. Therefore, modeling of a transformer in a steady state is very different from modeling in lightning transients. Modeling system elements in transients require a thorough understanding of the fundamental physics of the phenomenon and their impacts on the components. Transient conditions in power systems occur when the system is not operating in a steady-state condition anymore and transitions to a new steady-state point. However, this transition does not always land on an operable condition and rather may end up with a severely damaged system. Such cases may be lightning strikes to the ground adjacent to transmission lines or direct lightning strikes to high voltage substations. However, it must be mentioned that a major portion of transients in power systems is related to switching transients.

DOI: 10.1201/9781003255130-1

1

Circuit breakers and disconnectors isolate portions of the power grid. Circuit breakers are designed to operate when the system should be energized. Also, fuses and circuit breakers are designed to disconnect higher currents such as fault currents. During these transients, voltage and current transients occur in the order of microseconds to milliseconds.

1.2 CATEGORIES OF TRANSIENT PHENOMENON IN POWER SYSTEMS

Figure 1.1 illustrates a timetable of the transient phenomena in power systems. Transients on the left-hand side of the chart are mostly related to direct interactions of magnetic fields of inductances and electric fields of capacitors. These interactions are called electromagnetic phenomena. Transients on the right-hand side of the timetable are usually related to interactions between the stored mechanical energies in rotating machines and stored electrical energy in the grid. These phenomena are called electro-mechanical transients. The middle part of the timetable comprises both transients (i.e., electromagnetic and electro-mechanic). This part of the graph covers transient stability region. However, each phenomenon produces a waveform, which

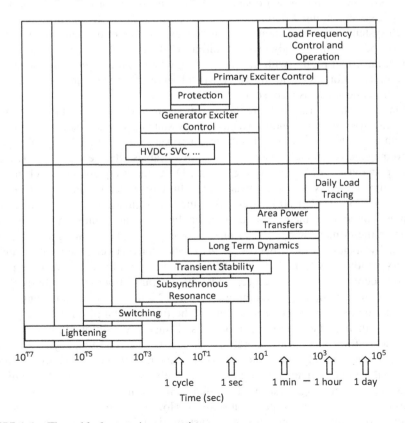

FIGURE 1.1 Timetable for transient cases in power systems.

can be expanded by Fourier's series. In any case, this transformation is a powerful tool for identifying the frequency spectrum of any phenomenon.

In general, lightning strikes cause the highest frequencies. For voltages less than or equal to 230 kV, they must be considered for insulation coordination. For system voltages of 400 kV and above, transient overvoltages due to line energization are used for insulation considerations. We will look deeper into these concepts in the following chapters.

Electromagnetic transients can be modeled by partial differential equations based on nodal and mesh analysis in circuits when they are energized by a specific input. This topic has been discussed in circuit theory textbooks and other references. Therefore, the reader of this book should be familiar with the terminologies and have a fundamental understanding of partial differential equations. Transient waveforms contain one or more oscillating modes. Therefore, the natural frequencies of these oscillations will identify the waveform. Accurate identification of these oscillations is directly related to the equivalent circuits that are modeled for the components of the system. There is no single model that can provide meaningful results for the entire time span of transient analysis. Therefore, depending on the considered final goals of the analysis, circuit components should be modeled. For example, the model of a transmission line for fault studies is different from the model of the same line for switching studies. Therefore, considering an appropriate time span is a key concept in analyzing power systems transients. For instance, lightning is the fastest transient phenomenon, so for modeling and analysis, nanosecond to microsecond time spans are appropriate in this case. In this time span, the changes in the power frequency waveforms (i.e., 50 or 60 Hz waves) are negligible and system controllers cannot detect and respond. On the contrary, stray capacitors and inductors associated with system components will produce the largest impact on response.

Switching transients require a time scale of microseconds to milliseconds. If the system recovery is the concern of the study, results should be provided for several cycles. Therefore, depending on the type of transient that has occurred in the system, a compatible study time scale has to be considered. For example, the time scale required for simulating a fast transient model with stray parameters (i.e., inductors and capacitors) would be different from a case study related to dynamics of controllers, which are performed on equivalent circuits. The simulation time step should be considered at least one-tenth of the smallest time constant in the system implementing reliable system parameters, and considering their alterations with frequency is as important as choosing a highly accurate solution methodology. Power system elements are modeled as lumped components such as available models for sources, capacitor banks, and reactors, or they are modeled as distributed components in case of overhead lines, underground, or submarine cables. Circuit models must be valid for frequencies in the range of 50 Hz to 100 kHz or even more. In this range of frequency, the parameters of the circuit such as impedances will change drastically with the change in frequency. Therefore, modeling in this domain should be frequency dependent besides requiring accommodation for nonlinear cases such as modeling surge arresters, breaker arcs, and transformers saturation. When dealing with transients, the appearance of the aforementioned nonlinear phenomenon is inevitable

TABLE 1.1

Transient Phenomenon and Frequency Spectrum

Phenomenon	Frequency Range
Ferroresonance	0.1 Hz to 1 kHz
Load rejection	0.1 Hz to 3 kHz
Fault clearing	50 Hz to 3 kHz
Overhead transmission line switching	0.1 Hz to 20 kHz
Transient recovery voltages	50 Hz to 100 kHz
Lightning-induced overvoltages	10 kHz to 3 MHz
Switching in gas insulated substations	100 kHz to 50 MHz

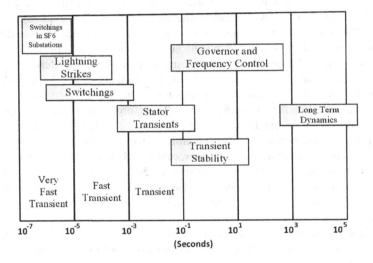

FIGURE 1.2   Time frame for power systems transient phenomenon.

considering the magnitude of overvoltages. Table 1.1 provides a list of common transients appearing in power systems along with their frequency spectrum.

Figure 1.2 categorizes power system transients into various sets. First, the cases related to switching in gas insulated substations (GIS) fall into the category of very fast transients. In this figure, the further we go along the line from left to right, the phenomenon will be slower and slower. The next set is fast transients for cases such as lightning strikes. Finally, there will be the category of transients; in this category, we have transients of the stator of the synchronous machines. It is worth noting that none of the abovementioned transients are detectable by protection systems. As the transient gets slower, the relays and protection systems start detecting and responding them. Therefore, in designing a safe and reliable power system, it is absolutely crucial to take into account the situations in which very fast and fast transients might arise.

1.3 TRANSIENTS AND THEIR EFFECT ON DESIGN AND OPERATION OF POWER SYSTEMS

As it is stated previously, power system protective devices operate in the range of power frequency, and therefore, they are not able to detect very fast and fast transients. Consequently, studying these cases in detail and with high accuracy is inevitable in designing power systems. The sanity of power systems should be maintained during and after occurrence of any transients. This should be meticulously considered in the design phases. Therefore, study of the possible transient situations will identify any system liabilities during the study, and the design can be altered with the system liability. Power system state can be viewed as a vector of n elements X in n dimension space. Let us assume that the normal operating point of system is X_0. The X_0 should be chosen in such a way that if disturbance happens and the system moves to a new operating point as X_1, this point is still a safe and reliable operating point in n dimensional space within limits stated by common standards (i.e., IEEE, IEC, etc.). Also, all the points of the trace from X_0 till X_1 must be within these limits. But what are the disturbances.

1.4 CATASTROPHIC CASES

We categorize the disturbance into four different groups as follows:

a. Overvoltages
b. Overcurrents
c. Distorted waveforms
d. Electromechanical transients

1.4.1 HIGH VOLTAGE TRANSIENTS

1.4.1.1 Required Studies

Safety and reliability are the most important factors to be considered in the design of any operable system. In the case of power systems, the number of blackouts will adversely affect the reliability indices of power systems. One of the most important causes of blackouts in the power system is overvoltages. In designing a system, there is always cost versus reliability consideration. This means that a designer can use the most durable insulation material that never fails throughout the system, and this will be the best solution for enhancing the system reliability and worst solution considering costs. On the other hand, ignoring the system reliability and loosening the regulation on system reliability indices may bring the best economical design along with the worst reliability considerations. In this latter approach, the system safety and dependability will be deteriorated. It should be noted for a \$/kWh energy production, the average of cost associated with system blackouts is approximately $40xa$. Obviously, none of the abovementioned solutions is a suitable approach. Therefore, the best solution considering the design of system insulation will be the result of accurate insulation coordination studies considering the amplitude of possible system overvoltages.

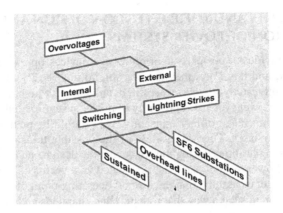

FIGURE 1.3   Overvoltage categories.

1.4.1.2   Causes of Overvoltages

Overvoltages can occur due to external or internal sources of disturbance in the system. An example of external disturbances is lightning and that of internal disturbance is transmission line switching. Figure 1.3 categorizes these disturbances. Table 1.2 illustrates disturbances and the resultant maximum overvoltage. The table lists the switching of overhead lines with rated voltages above 400 kV, which is of great importance due to the fact that the transient overvoltage is much higher than the tolerable voltage by the line. The results presented in this table will be discussed in the following chapters in detail.

1.4.1.3   Insulation Coordination and System Protection Against Transients

The abovementioned categories are important due to the fact that system elements show different behavior when dealing with different overvoltages. For example, an insulator designed for 400 kV system can tolerate different voltage levels in different conditions. In normal condition, it tolerates up to 245 kV line to ground voltage. This equipment can tolerate up to 650 kV in case of switching transients, and this level will move up to 900 kV when dealing with lightning strikes. Therefore, the same insulator shows three different behaviors depending on the type of the overvoltage

TABLE 1.2
Overvoltages and Maximum Transient Overvoltages

Phenomenon	Maximum Voltage (pu)	Duration Time Scale	Description
Lightning	2–3	Few microseconds	Important for V < 300 kV
Overhead lines switching	2–3	10 microseconds to a few milliseconds	Important for V > 300 kV
SF6 substations switching	2–3	Few microseconds	Important for V > 300 kV
Sustained	1.3–1.5	Period of power frequency	Occurrence due to protection system malfunction and waveform frequency is at power frequency

exposed on the system. As can be seen in Table 1.2, the time duration of the three mentioned phenomena and the overvoltages resulting from them are different; therefore, the elements show different levels of tolerance for each case. It is important to consider the following steps in order to prevent catastrophic events due to overvoltages. This process is also called insulation coordination.

Step 1 – Finding preventive ways such as shield wire installation on transmission line towers. Shield wire provides a direct path to ground for voltage surges due to lightning strikes.

Step 2 – Suppressing the amplitude of possible overvoltages, for example, by installing surge arresters.

Step 3 – Designing a high-performance and durable insulation system, which must withstand overvoltages.

1.5 OVERCURRENT TRANSIENTS

As mentioned before, the second category of power system disturbances is overcurrents. In power systems, faults are the main sources of high current transients. Studying high current transients is important when sizing the system circuit breakers and preventing mechanical and thermal stresses on system apparatuses.

1.5.1 CIRCUIT BREAKER DESIGN

Circuit breakers are chosen by the amount of fault current they can break and also the amount of transient recovery voltage they can tolerate after opening. Therefore, the system designer should have a good estimation of the system overcurrents.

1.5.2 MECHANICAL STRESS

Short circuit current will produce electromechanical forces due to the Lorentz law. These forces can displace and/or damage system equipment such as transformers. Also, transformer switching can cause high inrush currents and, as a result, high electromechanical forces.

1.5.3 THERMAL STRESS

Passing the short circuit current through system equipment containing loss elements will produce a large amount of heat and may lead to permanent damage of devices such as surge arresters.

1.6 ABNORMAL WAVEFORMS

1.6.1 IMPORTANCE OF THE STUDY

Delivering high-quality power to customers is one of the most important aspects in the design and operation of power systems. Electronic devices and power electronic drives are very sensitive to the quality of their input electric signals. Some of these

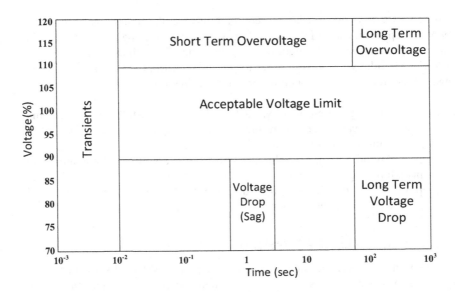

FIGURE 1.4   Categories of voltage disturbances.

equipment can tolerate large voltage sags and swell, but other equipment react to the
smallest power quality issue such as in PLLs and programmable logic controllers
(PLCs). In order to enhance the quality of power, the first step would be to understand the types and severity of the disturbances in the system. Some of the adverse
effects that have been observed in the system related to power quality issues are mal-
functions, aging of the equipment, increase in losses and consumption, faulty measurements, nuisance tripping of power electronic devices, and so many more. Types
of disturbances in waveform are summarized in the following section.Disturbances
in the system are categorized depending on the length of time that they affect the
system, and also, the magnitude of the transient overvoltage they cause. Figure 1.4
illustrates these categories. The first group is the phenomenon with time duration
less than half a cycle, which is called transients. The second group is the voltage
sags, which means more than 10% decrease in the voltage for a duration of 30 cycles
to 30 s. Voltage sags larger than 10%, which they last more than 1 min, are called
long-term voltage drops. Long voltage swell is considered when there is an increase
in voltage magnitude for more than 1 min and with magnitudes larger than 10%. For
time durations less than 1 min, this type of disturbance is called short-term voltage
swell. Beyond the specified situations, the system voltages lie in a normal range.
These subjects and concepts will not be discussed in more detail in this book, and
the books on power quality of power systems are good references for these studies.

1.7   ELECTROMECHANICAL TRANSIENTS

A mismatch between power generation of a generator and the power consumption of
loads connected to this generator results in electromechanical transients. The result
can be an increase or a decrease of generator rotor speed compared with its normal

rotation speed. The reason is usually a disturbance such as transmission line fault and subsequent line outage. This subject is known as power system transient stability and will not be studied in this book.

Also, other concepts such as low-frequency oscillations (LFOs) and sub-synchronous resonances (SSRs) are important subjects, known as the power system dynamics, which can be considered as electromechanical transients.

1.8   STUDY TECHNIQUES FOR TRANSIENT STUDIES

There are two common approaches in studying transients. The first one is using field tests on prototypes and the second one is modeling the phenomena in a computer-based platform and doing case simulations. Building a prototype is one of the methods that has been useful in design phases. Testing the prototype will provide the required data and help the designer in adjusting the design parameters. This approach is expensive and usually used for novel designs. In the operation phase, the field test is not recommended due to possible harmful overcurrent and overvoltages transients. The best methodology in understanding transient phenomenon is to model them. Both modeling and computer analysis can be used in the design phase and the operation phase. This will provide a safe environment along with accurate data generation.

1.8.1   Modeling and Simulation

a. Modeling
   In modeling a system, we need to know what kind of data we have available for the system. Also, we need to set clear expectations from the model. Therefore, a comprehensive description of the study phenomenon will be obtained depending on the data availability and expectations set. This procedure, that is, an abstract representation of the components and connections of a phenomenon is known as modeling. The available data and our expectations can be different; therefore, it is possible to have various models from one physical phenomenon. For example, in modeling a transformer for fast transient studies, we should estimate or determine the transformer transient model parameters using measurements or filed calculations, respectively, and our expectation is the ability to estimate the voltage stresses on transformer windings insulations. In this case, we need a detailed model of the windings, and terminal model of the transformer cannot be used. The same transformer can be modeled by its short circuit reactance for short circuit studies, and we do not need to use its detailed model.

b. Simulation
   The study and analysis performed on the abovementioned computer model, which is based on mathematical representations, is called simulation. Simulations are performed to tune or control a system by implementing its limitation and parameter variations. Also, in case of having non-analytical solution for a problem, we should use numerical methods and, as a result, simulations. Therefore, simulations are expected to result in a clear insight of the system vulnerabilities and capabilities.

There are two major categories in simulation:
- Deterministic Model Simulation
  In this type of simulation, no matter how many times we run the simulation, the outputs will be the same if the inputs and parameters are kept constant between simulations. In this case, all the variables are deterministic.
- Stochastic Model Simulation
  In this type of simulation, different runs of the same simulations will provide different results. In this case, the output of the simulation is also a random variable. This type of modeling needs accurate probability distribution functions for variables and inputs. In the end, the distribution of the output variables represents the most probable estimates and ranges of variations of the output variables. The Monte Carlo simulation comes under this group and can represent the effect of risk and uncertainty on models.

1.8.2 SIMULATIONS TOOLS

Simulations can be performed with a variety of concentrations on the following platforms:

a. Transient network analyzers
b. Analog computers
c. Digital computers
d. Hybrid computers
e. Hardware in the loop simulations

Digital simulation is the most common form. The main advantage of using digital simulations is the enhancements in the solution to algebraic and differential equations at discrete points. In the late 1960s, Professor H. W. Dommel in BPA (Bonneville Power Authority) proposed an applicable digital algorithm for studying transient phenomenon in power systems. This method was first implemented in EMTP (Electromagnetic Transients Program) software. This method works on the basis of differential equations, which were first proposed by Bergeron to solve transmission line equations. This method models transmission lines in a discrete time domain as traveling waves with constant parameters.

Bergeron model uses a fixed frequency model. Therefore, it is important to pay attention to the fact that although the Bergeron model can be used for transient simulations, the obtained results are meaningful for a fixed frequency of a specified system (50 or 60 Hz). This model is useful in studies such as relay testing or load-flow. For digital analysis, it is required to use discrete time scales and this may cause truncation errors and finally may lead to numerical instability. This has been improved by using trapezoidal methods in solving differential equations.

A number of simulation platforms are tabulated in Table 1.3. Simulation software programs such as PSCAD or EMTP are commonly used for any type of analysis wherein time domain simulations are required.

There are two basic streams of EMTP programs: the stream known simply as "EMTP" originates from the program development at BPA; other versions have been

TABLE 1.3
EMTP-Type Programs and Tools

Program	Provider	Website Info
EPRI/DCG EMTP	EPRI	www.emtp96.com/
ATP program		www.emtp.org/
MicroTran	Microtran Power Systems Analysis Corporation	www.microtran.com/
PSCAD/EMTDC	Manitoba HVDC Research Centre	www.hvdc.ca/
NETOMAC	Siemens	www.ev.siemens.de/en/pages/
NPLAN	BCP Busarello + Cott + Partner Inc.	
EMTAP	EDSA	www.edsa.com/
PowerFactory	DIgSILENT	www.digsilent.de/
Arene	Anhelco	www.anhelco.com/
Hypersim	IREQ (Real-time simulator)	www.ireq.ca/
RTDS	RTDS Technologies	https://www.rtds.com/
Transient Performance Advisor (TPA)	MPR (MATLAB-based)	www.mpr.com
Power System Toolbox	Cherry Tree (MATLAB-based)	www.eagle.ca/cherry/

TABLE 1.4
Less Popular Transient Analysis Software Platforms

Program	Provider	Website Info
ATOSEC5	University of Quebec at Trios Rivieres	https://www.uqtr.ca/english
Xtrans	Delft University of Technology	https://www.otnsystems.com/xtran
KREAN	The Norwegian University of Science and Technology	www.elkraft.ntnu.no/sie10aj/Krean1990.pdf
Power Systems	MathWorks (MATLAB-based)	www.mathworks.com/products/
Pouya	TOM Industrial Consulter	https://www.tomcad.com/
Blockset	TransEnergie Technologies	www.transenergie-tech.com/en/
SABER	Avant (formerly Analogy Inc.)	www.analogy.com/
SIMSEN	Swiss Federal Institute of Technology	https://www.siemens.com

written from scratch. The EMTP-ATP and MT-EMTP programs, for example, are based on the original BPA and DCG-EMTP versions. The alternate stream of EMTP-type programs and tools may use new numerical methods and modeling approaches, and provide significantly improved capabilities and numerical performances. Examples of this alternate stream include RTDS Technologies, PSCAD-EMTDC, EMTP-RV, MT-EMTP, EMTP-ATP, eMEGAsim, and HYPERsim from Opal-RT Technologies (Table 1.4).

It is worth noting that EMTP-type analyses are not substitutes for power flow and steady-state analysis. Rather, they are used to study electromagnetic transients in power systems. Although the EMTP-type software programs are not directly used for harmonic analysis, they can provide accurate results in harmonic power flow analysis and multi-machine transient stability analysis. In this book, the intent is to utilize PSCAD software and demonstrate the concepts in a hands-on way using the mentioned platform. This software provides time domain solutions that are most suitable when dealing with fast transients. Many of the complicated models that are discussed in this book and other references exist in PSCAD library. So, the reader of this book will have the opportunity to use the prebuilt models in PSCAD and examine case studies and understand the effects of transient cases in the system.

1.8.3 Important Considerations for Transient Analysis

Modeling a transient phenomenon depends not only on the model that is used but also on the platform wherein the simulation is going to be performed. We are summarizing some of the important rules in choosing appropriate models and practical simulation tools here.

a. Choosing a proper geometrical zone (dimensions) of study is one of the most important points in the study. When dealing with very large frequencies, the zone of the study should be very small.

b. In transient analysis, we need to focus on smaller pieces of the system with more accurate models. Adding more components to the study does not necessarily mean increasing the study accuracy. In many cases, adding more components may lead to inaccurate result generation. Also, the higher the system component numbers, the longer will be the simulation time.

c. Considering an accurate representation of system losses, in some cases such as Ferroresonace and capacitor banks switching, play an important role in determining the magnitude of overvoltages.

d. The ideal components model can be considered if the system under study is becoming too large to handle.

e. Sensitivity analysis plays an important role in cases where accurate values are not available for several parameters in the system. This type of analysis will identify which parameter(s) play a more important role in the system behavior. Input data and estimations in numerical solutions are the major components of error accumulation in the any study.

f. If there is frequency-dependent behavior of some devices, the study should be done in frequency domain, and if the system is dealing with nonlinear elements, time domain analysis will be suitable. In case of having variations both in frequency domain and time domain, z transformation will be required. This transformation needs numerous inputs from power system apparatuses and usually is not practical to implement.

1.9   AN INTRODUCTION TO PSCAD/EMTDC

The book uses PSCAD (power systems computer-aided design) software to illustrate the transient cases discussed in this book. The analysis can be mathematically intensive, but the software support provides a hands-on environment to the reader to examine each case and understand the mathematical calculations in a clearer way. This software provides a visual, user-friendly environment to assess power system transients discussed in the book. Audience interested in studying and understanding power systems transients is equipped with base models in this book that can be expanded to be used in studies and engineering projects related to system transients.

Problems

1.1 What are the different methodologies in analyzing transients in power systems? Specify pros and cons of each method.

1.2 In the following circuit, discuss the effect of SVC (Static VAr Compensator) on the voltage as labeled $v_2(t)$ when the breaker k closes at $t = 0$ in the interval $0 < t < 100$ μs.

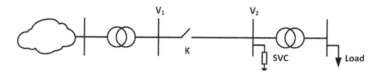

FIGURE P1.2   Switching in a power system with a static VAr compensator.

BIBLIOGRAPHY

1. Lou van der Sluis, *Transients in Power Systems*, John Wiley & Sons, 2001.
2. Juan A. Martinez-Velasco, *Transient Analysis of Power Systems: A Practical Approach*, John Wiley & Sons, 2019.
3. Neville Watson and Jos Arrillaga, *Power Systems Electromagnetic Transients Simulation*, London, United Kingdom: The Institution of Engineering and Technology, 2007.
4. Juan A. Martinez-Velasco, *Power System Transients: Parameter Determination*, the United States of America: Taylor & Francis Group, 2010.
5. J. C. Das, *Transients in Electrical Systems: Analysis, Recognition, and Mitigation*, the United States of America: McGraw Hill, 2010.
6. N. Mohan, W. P. Robbins, T. M. Undeland, R. Nilssen, and O. Mo, "Simulation of Power Electronic and Motion Control Systems – An Overview", *Proceedings of the IEEE*, Vol. 82 (8), 1994, pp. 1287–1302.
7. Amir Heidary, Hamid Radmanesh, and G. B. Gharehpetian, "Effect of Series Resonance LC Tank on Mitigation of Fault Current in Radial Distribution Networks", *Indian Journal of Science and Technology*, Vol. 9, Issue 7, February 2016, pp. 1–7.
8. A. Abbasi, S. H. Fathi, G. B. Gharehpatian, A. Gholami, and H. R. Abbasi, "Voltage Transformer Ferroresonance Analysis using Multiple Scales Method and Chaos Theory", *Complexity*, Vol. 18, No. 6, July/August 2013, pp. 34–45.
9. H. Radmanesh, G. B. Gharehpetian, and S. H. Fathi "Ferroresonance of Power Transformers Considering Nonlinear Core Losses and Metal Oxide Surge Arrester Effects", *Electric Power Components and Systems*, Vol. 40, issue 5, May 2012, pp. 463–479.
10. A. Ajami, S. H. Hosseini, and G. B. Gharehpetian, "Modeling and Controlling of UPFC for Power System Transient Studies", *ECTI Transactions on Electrical Engineering, Electronics and Communications*, Vol. 5, No. 2, August 2007.

11. M. H. Nazemi and G. B. Gharehpetian, "Reduced Order Model of Power Transformers for Power System Transient Studies", *Modern Electric Power Systems (MEPS'06)*, September 6–8, 2006, Wroclaw, Poland.

12. H. Alipour and G. B. Gharehpetian, *Power System Transient Studies Using EMTP*, Book, Amirkabir University Publication, Tehran, Iran, 1st Edition 1999, 2nd Edition, 2001, ISBN: 964-463-051-3 (in Persian).

13. G. B. Gharehpetian and A. Ghanizadeh, *Using EMTP-RV*, Book, Nahr Danesh Publication, Tehran, Iran, 1st Edition 2012, 2nd Edition, 2013, ISBN: 978-600-5714-13-5 (in Persian).

14. P. Karimifard and G. B. Gharehpetian, "On-Line Parameter Estimation of Transformers Using Power System Transients", *15th International Symposium on High Voltage Engineering, ISH 2007*, August 27–31, 2007, Ljubljana, Slovenia.

15. https://www.pscad.com/knowledge-base/topic-36/v-

2 Traveling Waves

A power system is a vast entity. In case of steady-state analysis of the system, components are modeled as lumped elements. This is due to the fact that the wavelengths associated with sinusoidal currents and voltages are considerably larger than the physical dimensions of system components. The transients utilizing lumped sum models will render inaccurate results, as the time taken for the electromagnetics wave to travel through system elements needs to be considered. Obviously, using π models cannot provide the traveling time of electromagnetic waves. In modeling an overhead line with sections of π lines, we assume that the characteristics of the associated electric field and magnetic field are jammed in one capacitor and one inductor, respectively. Figure 2.1 shows a source switching to a line. When the switch S closes, current passes through the inductance L_1 and later charges the capacitor C_1. The extra charge on C_1 will provide a voltage difference across L_2, causing current flow at that point. Next, the current will start charging C_2, and consequently, the voltage difference across L_3 will make a current flow. What is explained here clarifies that a disturbance on one side of the power system network will not be appear on the other side immediately. Experience has shown that the explained process in the π model is not the result of connecting the source to the line but rather the elements cause the time lapse in traveling wave from the source to the end of the line. Also, it is possible to alter the lumped sum parameters in the system and replace them with time domain differential equations. This method will not consider a change in position. Therefore, the traveling time of the electromagnetic waves cannot be obtained when using lumped sum elements. The only time that the mentioned process in analyzing the system transients work is when the dimensions of components are much less than the system wavelengths. Knowing the physical dimensions of the lines, in order to resolve the issue in finding the time it takes for electromagnetic wave to travel, we model the transmission line with distributed elements. Evenly distributed capacitors

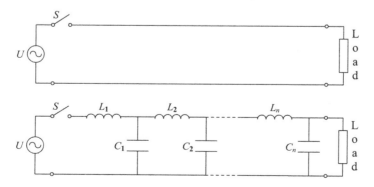

FIGURE 2.1 Lumped sum element illustration of a single-phase line.

DOI: 10.1201/9781003255130-2

and inductors along a transmission line will be considered in this case. It is worth noting that currents and voltages along a transmission line will have different values in different places for a considered time mark.

2.1 TRANSMISSION LINE EQUATIONS

Figure 2.2 illustrates an element of a single-phase transmission line with the length Δx. The capacitance between the two lines is $\Delta C = C'\Delta x$. C' is the capacitance in per unit length. The amount of inductance for the same portion of the line is $\Delta L = L'\Delta x$ and L' is the amount of inductance in per unit length. The amount of resistance in the same portion of line is $\Delta R = R'\Delta x$, where R' is the resistance in per unit length of the line. The conductance can be modeled across the capacitor as shown in the figure, which is $\Delta G = G'\Delta x$ and G' is the conductivity in per unit length.

Equation (2.1) shows the aforementioned quantities.

$$L' = \frac{dl}{dx}, \quad C' = \frac{dC}{dx}$$
$$G' = \frac{dG}{dx}, \quad R' = \frac{dR}{dx} \tag{2.1}$$

Now we begin with the problem of establishing partial differential equations. The next step will be to expand the equations and find the appropriate equations for the transmission line. As it can be seen in Figure 2.2, at the position x, which the beginning of the line, the voltage is $u(x)$ and current is $i(x)$. At the end of this portion of line with the position of $x + \Delta x$, $u(x + \Delta x)$ is the voltage and $i(x + \Delta x)$ is the current. As can be seen, voltage and current are functions of time and position. Therefore, the voltage drop across this piece of line can be written as Equation (2.2).

$$\Delta u = L'\Delta x \frac{\partial i(x,t)}{\partial t} + R'\Delta x i(x,t) \tag{2.2}$$

$$\Delta u = u(x,t) - u(x + \Delta x,t) \tag{2.3}$$

Therefore,

$$u(x + \Delta x,t) = u(x,t) - R'\Delta x i(x,t) - L'\Delta x \frac{\partial i(x,t)}{\partial t} \tag{2.4}$$

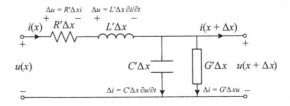

FIGURE 2.2 Transmission line model with Δx length.

$$\frac{u(x+\Delta x,t)-u(x,t)}{\Delta x} = -R'i(x,t)-L'\frac{\partial i(x,t)}{\partial t} \tag{2.5}$$

Now, if we find the limit of Equation (2.5) when Δx goes to zero, Equation (2.6) will be obtained.

$$-\frac{\partial u(x,t)}{\partial x} = R'i(x,t)+L'\frac{\partial i(x,t)}{\partial t} \tag{2.6}$$

We can also write the same equations for the current, as follows:

$$-\frac{\partial i(x,t)}{\partial x} = G'u(x,t)+C'\frac{\partial u(x,t)}{\partial t} \tag{2.7}$$

If we do a partial derivative with respect to x and t, respectively, from both sides of Equations (2.6) and (2.7), we can find Equations (2.8) and (2.9).

$$-\frac{\partial^2 u}{\partial x^2} = R'\frac{\partial i}{\partial x}+L'\frac{\partial^2 i}{\partial x\partial t} \tag{2.8}$$

$$-\frac{\partial^2 i}{\partial x\partial t} = G'\frac{\partial u}{\partial t}+C'\frac{\partial^2 u}{\partial t^2} \tag{2.9}$$

Substituting Equation (2.7) in (2.8), we will have (2.10).

$$\frac{\partial^2 u}{\partial x^2} = R'G'u+R'C'\frac{\partial u}{\partial t}-L'\frac{\partial^2 i}{\partial x\partial t} \tag{2.10}$$

Also, using Equation (2.9), we can rewrite Equation (2.10) as Equation (2.11).

$$\frac{\partial^2 u}{\partial x^2} = R'G'u+(R'C'+L'G')\frac{\partial u}{\partial t}+L'C'\frac{\partial^2 u}{\partial t^2} \tag{2.11}$$

We can apply the same process for finding the equations of currents. Equation (2.12) is obtained by taking a partial derivative from both sides of Equations (2.7) and (2.6) with respect to x and t, respectively. We will get Equations (2.12) and (2.13), respectively. Substituting Equation (2.6) into Equation (2.12), will give Equation (2.14).

$$-\frac{\partial^2 i}{\partial x^2} = G'\frac{\partial u}{\partial x}+C'\frac{\partial^2 u}{\partial x\partial t} \tag{2.12}$$

$$-\frac{\partial^2 u}{\partial x\partial t} = R'\frac{\partial i}{\partial t}+L'\frac{\partial^2 i}{\partial t^2} \tag{2.13}$$

$$\frac{\partial^2 i}{\partial x^2} = R'G'i + L'G'\frac{\partial i}{\partial t} - C'\frac{\partial^2 u}{\partial x \partial t} \tag{2.14}$$

Now combining Equations (2.13) and (2.14), Equation (2.15) will be obtained.

$$\frac{\partial^2 i}{\partial x^2} = R'G'i + (R'C' + L'G')\frac{\partial i}{\partial t} + L'C'\frac{\partial^2 i}{\partial t^2} \tag{2.15}$$

Now considering the operand P as $\dfrac{d}{dt}$, we can change Equations (2.11) and (2.15) into Equation (2.17).

$$\begin{cases} \dfrac{\partial^2 u}{\partial x^2} = (R' + L'P)(G' + C'P)u \\[2mm] \dfrac{\partial^2 i}{\partial x^2} = (R' + L'P)(G' + C'P)i \end{cases} \tag{2.17}$$

In order to simplify Equation (2.17), we consider the following substitutions as in Equation (2.18).

$$\begin{cases} y' = G' + C'P \\ z' = R' + L'P \end{cases} \tag{2.18}$$

Therefore, Equation (2.17) can be written as:

$$\begin{cases} \dfrac{\partial^2 u}{\partial x^2} = z'y'u \\[2mm] \dfrac{\partial^2 i}{\partial x^2} = z'y'i \end{cases} \tag{2.19}$$

In the above equations, we can define the propagation constant as follows:

$$\gamma = \sqrt{z'y'} \tag{2.20}$$

In order to solve the partial differential equations in (2.19), we form the characteristic equation as (2.21).

$$m^2 - \gamma^2 = 0 \Rightarrow m = \pm\gamma \tag{2.21}$$

Therefore, the general solution can be written as Equation (2.22).

$$u = f_1 e^{\gamma x} + f_2 e^{-\gamma x} \tag{2.22}$$

Considering the fact that f_1 and f_2 are functions of time, Equation (2.22) can be written as Equation (2.23).

$$u = f_1(t)e^{\gamma x} + f_2(t)e^{-\gamma x} \tag{2.23}$$

It is worth noting that u is a function of time and position. Applying the operand P into Equation (2.6) and considering Equation (2.18), we can find Equations (2.24) and (2.25).

$$\frac{\partial u}{\partial x} = -z'i \tag{2.24}$$

$$i = -\frac{1}{z'}\frac{\partial}{\partial x}(f_1(t)e^{\gamma x} + f_2(t)e^{-\gamma x}) \tag{2.25}$$

Now, we define a new parameter and will show that parameter with Z_o.

This parameter is called the line characteristic impedance or the line surge impedance. In fact, the characteristic impedance of the line is the impedance that the line shows in transient situations as shown in Equation (2.26).

$$Z_o = \sqrt{\frac{z'}{y'}} \tag{2.26}$$

Also, the characteristic admittance can be written as Equation (2.27).

$$Y_o = \frac{1}{Z_o} = \sqrt{\frac{y'}{z'}} \tag{2.27}$$

So far, we have managed to write the equations governing voltage and current in the transmission line. While working on those equations, we introduced the characteristic impedance of the transmission line. Next step will be to assess the response of the transmission in different conditions.

2.1.1 LOSSLESS LINES

In this condition, parameters R' and G' are set to be zero. Therefore, we will find Equation (2.28).

$$\begin{cases} \gamma = \sqrt{z'y'} = P\sqrt{L'C'} \\ Z_o = \sqrt{\dfrac{z'}{y'}} = \sqrt{\dfrac{L'}{C'}} \end{cases} \tag{2.28}$$

The speed of the propagation wave can be written as Equation (2.29).

$$v = \frac{1}{\sqrt{L'C'}} = \frac{P}{P\sqrt{L'C'}} = \frac{P}{\gamma} \tag{2.29}$$

Therefore, we have:

$$u = f_1(t)e^{\frac{xP}{v}} + f_2(t)e^{-\frac{xP}{v}} \tag{2.30}$$

$$i = -\frac{1}{Z_o}\frac{\partial}{\partial x}\left[f_1(t)e^{\frac{xP}{v}} + f_2(t)e^{-\frac{xP}{v}} \right] \tag{2.31}$$

Equations (2.30) and (2.31) are obtained by substituting Equation (2.29) into Equations (2.23) and (2.25). Using the following Taylor expansion, we will prove that the traveling waves will move along the x-axis to the left and right.

$$f(t+a) = f(t) + af(t) + \frac{a^2}{2!}f'(t) + \cdots = \left(1 + aP + \frac{a^2}{2!}P^2 + \cdots \right)f(t) = e^{aP}f(t) \tag{2.32}$$

Using Equation (2.32), we will rewrite Equations (2.30) and (2.31) as Equation (2.33):

$$e^{\frac{xP}{v}}f_1(t) = f_1\left(t + \frac{x}{v} \right) \text{ or } f_3(x+vt) \tag{2.33}$$

We can rewrite Equations (2.30) and (2.31) as it is shown in Equation (2.34).

$$\begin{cases} u(x,t) = f_3(x+vt) + f_4(x-vt) \\ i(x,t) = -\dfrac{1}{Z_o}[f_3(x+vt) - f_4(x-vt)] \end{cases} \tag{2.34}$$

Considering Figure 2.3, the function $f(x - vt)$ at $t = 0$ will be $f(x)$, as can be written in Equation (2.35).

$$f(x - vt)\big|_{t=0} = f(x) \tag{2.35}$$

For the position $x = a$, the amount of the function is as follows:

$$f(x)\big|_{x=a} = f(a) \tag{2.36}$$

Now, at $t = \tau$, the position will be $x = a + v\tau$, and the amount of the function can be determined, as below:

$$f(x - vt)\big|_{t=\tau} = f(x - vt)\big|_{x=a+v\tau} = f(a) \tag{2.37}$$

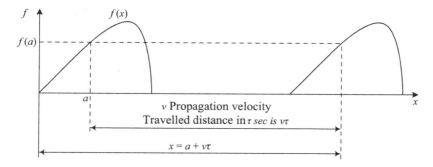

FIGURE 2.3 Wave propagation on the positive side of x-axis.

Therefore, it can be said that the amount of the function has not been changed during time. The same can be said for the all the points of the curve $f(x - vt)$. As a result, the waveform preserves its shape, when it moves toward the positive side of the x-axis. The function $f(x + vt)$ is the same as the above illustrated function where v is changed to $-v$. In other words, $f(x + vt)$ is the wave that is traveling toward the negative side of the x-axis. Therefore, considering (2-34) at $t = 0$, the voltage waveform will have two parts. One part travels to the positive side of the axis and the other part travels to the negative side of the x-axis. The same is valid for the current as well.

Based on Equation (2.38),

$$v = \frac{dx}{dt} \Rightarrow \frac{dx}{dt} - v = 0 \Rightarrow x - vt = \text{constant} \tag{2.38}$$

we can predict that the arguments of functions f_3 and f_4 are always constant. Therefore, the shape and magnitude of the wave in the process of traveling will not be altered.

2.1.2 LOSSY LINES

a. Analysis

Considering the definition of rotating phasors, we have:

$$x(t) = A_m \cos(\omega t + \phi) \Leftrightarrow A = A_m e^{j\phi} \tag{2.39}$$

$$x(t) = \text{Re}\left\{A_m e^{j\phi}\right\} \Leftrightarrow A_m e^{j\phi} \tag{2.40}$$

where $A_m e^{j\phi}$ is the rotating phasor. In order to solve the partial differential equations, we assume Equations (2.41)

$$\begin{cases} u(x,t) = \text{Re}\left\{\overline{u}(x)e^{j\omega t}\right\} \\ i(x,t) = \text{Re}\left\{\overline{i}(x)e^{j\omega t}\right\} \end{cases} \tag{2.41}$$

Replacing Equations (2.41) into (2.11), we will have Equations (2.42)–(2.44).

$$\frac{\partial^2 u}{\partial x^2} = R'G'\overline{u} + j\omega(R'C' + L'G')\overline{u} - \omega^2 L'C'\overline{u} \tag{2.42}$$

$$\frac{\partial^2 u}{\partial x^2} = (R' + j\omega L')(G' + j\omega C')\overline{u} = z'y'\overline{u} \tag{2.43}$$

$$\frac{d^2 u}{dx^2} = z'y'\overline{u} \tag{2.44}$$

It is worth noting that in z' and y', the operand P is replaced by $j\omega$. In order to solve the abovementioned partial differential equation, let us form the characteristic equation as in Equation (2.45).

$$m^2 - \gamma^2 = 0 \Rightarrow m = \pm\sqrt{z'y'} \tag{2.45}$$

Then, we have:

$$\overline{u}(x) = \overline{u}_1 e^{-\gamma x} + \overline{u}_2 e^{+\gamma x} \tag{2.46}$$

We know:

$$-\frac{d\overline{u}}{dx} = (R' + j\omega L')\overline{i} = z'\overline{i} \tag{2.47}$$

Therefore, we have:

$$\overline{i} = \frac{\overline{u}_1}{z'} e^{-\gamma x} - \frac{\overline{u}_2}{z'} e^{+\gamma x} \tag{2.48}$$

$$i(x) = \gamma\left(\frac{\overline{u}_1}{z'} e^{-\gamma x} - \frac{\overline{u}_2}{z'} e^{+\gamma x}\right) \tag{2.49}$$

Considering the fact that z' and Z_0 are related through $Z_0 = \dfrac{z'}{\gamma}$, it is possible

to rewrite Equation (2.49) as Equation (2.50) or (2.51)

$$\overline{i}(x) = \frac{\overline{u}_1(x)}{Z_o} - \frac{\overline{u}_2(x)}{Z_o} \tag{2.50}$$

$$\begin{cases} \bar{i_1}(x) = \dfrac{\bar{u}_1(x)}{Z_o} \\[3mm] \bar{i_2}(x) = \dfrac{\bar{u}_2(x)}{Z_o} \end{cases} \tag{2.51}$$

We expand Equation (2.51) to Equation (2.52):

$$\begin{cases} \bar{i}(x) = i_f + i_r \\[2mm] \bar{u}(x) = u_f + u_r \\[2mm] i_f = \dfrac{u_f}{Z_o} \\[3mm] i_r = -\dfrac{u_r}{Z_o} \end{cases} \tag{2.52}$$

Considering the fact that using Fourier transformation, any waveform can be decomposed into frequency elements, therefore, a transient waveform can be presented with its frequency elements. Each frequency element or sine wave can be used for a sinusoidal state study.

b. Important findings

We assume the phasors at the beginning of the line are as shown in Equation (2.53).

$$\begin{cases} \bar{u}(x = 0) = \bar{u}_o \\[2mm] \bar{i}(x = 0) = \bar{i}_o \end{cases} \tag{2.53}$$

Considering Equation (2.53) and using Equations (2.46) and (2.49), we obtain Equations (2.54) and (2.55)

$$\begin{cases} \bar{u}_o = \bar{u}_1 + \bar{u}_2 \\[2mm] \bar{i}_o = \left(\dfrac{\bar{u}_1 e^{-\gamma x}}{Z_o} - \dfrac{\bar{u}_2 e^{\gamma x}}{Z_o} \right)\Bigg|_{x=0} = \dfrac{\bar{u}_1}{Z_o} - \dfrac{\bar{u}_2}{Z_o} \end{cases} \tag{2.54}$$

$$\begin{cases} \bar{u}_1 = \dfrac{\bar{u}_o + Z_o \bar{i}_o}{2} \\[3mm] \bar{u}_2 = \dfrac{\bar{u}_o - Z_o \bar{i}_o}{2} \end{cases} \tag{2.55}$$

Therefore, we will have

$$\begin{cases} \bar{u}(x) = \dfrac{\bar{u}_o}{2}(e^{\gamma x} + e^{-\gamma x}) - \dfrac{Z_o \bar{i}_o}{2}(e^{\gamma x} - e^{-\gamma x}) \\[3mm] \bar{i}(x) = \dfrac{\bar{i}_o}{2}(e^{\gamma x} + e^{-\gamma x}) - \dfrac{\bar{u}_o}{2Z_o}(e^{\gamma x} - e^{-\gamma x}) \end{cases} \tag{2.56}$$

It can be seen that the initial conditions are considered in Equation (2.56). Also, Equations (2.56) are satisfying Equation (2.44). We can change Equation (2.56) to Equation (2.57).

$$\begin{cases} \bar{u}(x) = \bar{u}_o \cosh \gamma x - Z_o \bar{i}_o \sinh \gamma x \\ \bar{i}(x) = \bar{i}_o \cosh \gamma x - \dfrac{\bar{u}_o}{Z_o} \sinh \gamma x \end{cases} \tag{2.57}$$

where γ is the propagation parameter. This parameter is dependent on frequency, and therefore, is not a constant, because in these studies, frequency is not constant.

$$\gamma = \sqrt{(R' + j\omega L')(G' + j\omega C')} = \alpha + j\beta \tag{2.58}$$

We can write α and β using the characteristics of the line.

$$\alpha = \sqrt{\frac{1}{2}(R'G' - \omega^2 L'C') + \frac{1}{2}\sqrt{(R'^2 + \omega^2 L'^2)(G'^2 + j\omega^2 C'^2)}} \tag{2.59}$$

$$\beta = \sqrt{\frac{1}{2}(\omega^2 L'C' - R'G') + \frac{1}{2}\sqrt{(R'^2 + \omega^2 L'^2)(G'^2 + j\omega^2 C'^2)}} \tag{2.60}$$

where α and β are, respectively, called attenuation parameter and phase parameter. If we consider ω the power frequency, Equations (2.57) will be the same steady-state voltage and current equations at position x on the line, knowing the fact that voltage and current at $x = 0$ are given in Equation (2.53). These equations have been presented as line steady-state equations in power system elementary books.

c. Damping effect
   We know,

$$u(x,t) = Re\{\bar{u}(x)e^{j\omega t}\} \tag{2.61a}$$

Which,

$$\bar{u}(x) = \bar{u}_1 e^{-\gamma x} + \bar{u}_2 e^{\gamma x} \tag{2.61b}$$

Therefore,

$$u(x,t) = Re\{(\bar{u}_1 e^{-\gamma x} + \bar{u}_2 e^{\gamma x})e^{j\omega t}\} \tag{2.61c}$$

Knoawing $\gamma = \alpha + j\beta$, we will have:

$$u(x,t) = Re\{\bar{u}_1 e^{-\alpha x} e^{j(-\beta x + \omega t)} + \bar{u}_2 e^{+\alpha x} e^{j(\beta x + \omega t)}\} = u_1(x,t) + u_2(x,t) \tag{2.62}$$

Equation (2.62) is showing a wave that has two parts. One wave is traveling to the right-hand side of the x-axis and the other wave to the left-hand side of the x-axis. The effect of damping will cause the wave to attenuate as it travels along the x-axis.

In Equation (2.63),

$$u_1(x,t) = Re\left\{\overline{u}_1 e^{-\alpha x} e^{j(-\beta x + \omega t)}\right\} \tag{2.63}$$

if t increases with Δt and x increases with Δx, and considering Equation (2.64), $(-\beta x + \omega t)$ will not be a time-dependent function. We are proving this in Equations (2.65) and (2.66)

$$\Delta x = v\Delta t = \frac{\omega}{\beta}\Delta t \tag{2.64}$$

$$[-\beta(x + \Delta x) + \omega(t + \Delta t)] = (-\beta x - \beta \Delta x + \omega t + \omega \Delta t) = -\beta x + \omega t \tag{2.65}$$

Considering this point that this wave travels with the speed of v,

$$v = \frac{\Delta x}{\Delta t} = \frac{\omega}{\beta} \tag{2.66}$$

Therefore,

$$u_1(x,t) = Re\left\{\overline{u}_1 e^{-\alpha x} e^{-j\beta\left(x - \frac{\omega}{\beta}t\right)}\right\} = Re\left\{\overline{u}_1 e^{-\alpha x} e^{-j\beta(x-vt)}\right\} \tag{2.67}$$

Now, if we consider $e^{-j\beta(x-vt)} = f(x-vt)$, will render Equation (2.68):

$$u_1(x,t) = Re\{\overline{u}_1 e^{-\alpha x} f(x-vt)\} \tag{2.68}$$

which shows that the wave is moving along the positive side of the x-axis; also, the function $e^{-\alpha x}$ attenuates the waveform as it travels along the x-axis. Now, considering the wave that travels to the left side direction of the x-axis, we will have Equation (2.69). This wave approaches the sending end of the line.

$$u_2(x,t) = Re\{\overline{u}_2 e^{\alpha x} f(x+vt)\} \tag{2.69}$$

While,

$$v = -\frac{\omega}{\beta} \tag{2.70}$$

The wave travels toward the sending end of the line, and with time the value of x decreases, therefore the component $e^{\alpha x}$ as a multiplier attenuates the signal.

In summary, we can write Equation (2.71) as the waves traveling to the right side and left side of the x-axis will damp out completely as the time goes.

$$\lim_{x\to+\infty} u_1(x,t) = 0, \qquad \lim_{x\to-\infty} u_2(x,t) = 0 \qquad (2.71)$$

2.2 REFLECTION RULES FOR SINUSOIDAL WAVES

2.2.1 TRANSMISSION LINE REFLECTION

There is a fixed relationship between current and voltage, while an electromagnetic wave is propagating along a line with a known characteristic impedance. When the wave approaches a discontinuity such as an open circuit, a short circuit or where the new characteristic impedance is $Z = \left(\dfrac{L}{C}\right)^{1/2}$ (e.g., a place where the line gets connected to a cable or a transformer), interesting conditions will happen. Due to the existence of a mismatch between the line impedance and the characteristic impedance, mismatch between the voltage and current waves should happen. In encountering a discontinuity, some of the energy of the wave passes through and some will get reflected and started moving in the opposite direction. At the discontinuity point, the waveforms of voltage and current are identical. Ignoring the system losses, the amount of energy will not change encountering the discontinuity. Figure 2.4 shows a condition wherein a transmission line gets connected to an underground cable. In order to pursue simplification, both the line and the cable are considered to be lossless (R = G = 0). Characteristic impedances for line and underground cable are formulated in Equations (2.72) and (2.73).

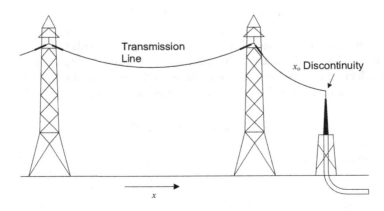

FIGURE 2.4 Transmission line terminated to an underground cable.

$$Z_L = \sqrt{\frac{L_L}{C_L}} \qquad (2.72)$$

$$Z_C = \sqrt{\frac{L_C}{C_C}} \qquad (2.73)$$

The incident wave is the one that moves along the positive x-axis. This positive direction in Figure 2.4 is from left to right.

2.2.2 END OF THE LINE REFLECTION

$$Z_L$$

At any points on the line, there will be two travelling waves. One of the waves is traveling to the right-hand side and the other wave is traveling to the left-hand side. According to Equation (2.61c), at each point, the voltage wave will be the equivalent of the right wave and left wave and the same is valid for the current wave. Figure 2.5 illustrates the parameters that are going to be used in this section.

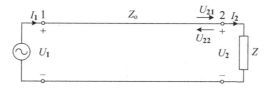

FIGURE 2.5 Wave reflection, transmission line is terminated to impedance Z.

In this figure:

U_2 Voltage at point 2
U_{21} Voltage traveling wave toward positive x-axis at point 2
U_{22} Voltage traveling wave toward negative x-axis at point 2
I_2 Current wave at point 2
I_{21} Current traveling wave toward positive x-axis at point 2
I_{22} Current traveling wave toward negative x-axis at point 2

At point 2, we can write Equation (2.74)

$$\begin{cases} U_2 = U_{21} + U_{22} \\ I_2 = I_{21} + I_{22} \end{cases} \qquad (2.74)$$

On the other hand, in a lumped sum circuit, we have:

$$U_2 = ZI_2 \qquad (2.75)$$

Considering Equation (2.52), we can write Equation (2.76)

$$\begin{cases} U_{21} = Z_o I_{21} \\ U_{22} = -Z_o I_{22} \end{cases} \qquad (2.76)$$

Therefore:

$$\begin{cases} U_2 = U_{21} + U_{22} \\ \dfrac{U_2}{Z} = \dfrac{U_{21}}{Z_o} - \dfrac{U_{22}}{Z_o} \end{cases} \qquad (2.77)$$

Also, this last equation can be written as:

$$\begin{cases} U_2 = U_{21} + U_{22} \\ U_2 \dfrac{Z_o}{Z} = U_{21} - U_{22} \end{cases} \qquad (2.78)$$

Therefore:

$$U_2 = \frac{2Z}{Z + Z_o} U_{21} \qquad (2.79)$$

Equation (2.79) will illustrate the fact that the wave entering a new environment can be obtained using the terminated impedance and the line characteristic impedance.

According to Figure 2.6, there is resemblance to the light reflection due to the change in characteristic impedance.

As it is explained above, U_{22} is the voltage traveling wave going toward the negative x-axis. Equation (2.80) shows this voltage wave.

$$U_{22} = U_2 - U_{21} = \frac{Z - Z_o}{Z + Z_o} U_{21} \qquad (2.80)$$

According to Equations (2.79) and (2.80), the reflecting and passing waves are both constant multipliers of the incoming wave. These multipliers can be defined as:

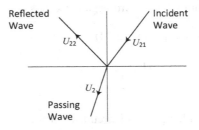

FIGURE 2.6   Reflection and passing of a wave at the border of two environments.

$$\begin{cases} b = \dfrac{2Z}{Z+Z_o} \\[4mm] r = \dfrac{Z-Z_o}{Z+Z_o} \end{cases} \qquad (2.81)$$

In Equation (2.81), b is the break coefficient and r is the reflect coefficient. Equation (2.82) can be concluded from the above equations.

$$b - r = 1 \qquad (2.82)$$

We will continue the subject with special cases.

a. Line terminated to characteristic impedance Z_0

   Let us assume that the terminating impedance at the end of the line is the same as the line characteristic impedance $Z = Z_0$. Therefore, the reflection coefficient will be zero and the break coefficient will be 1. Using Equations (2.79) and (2.80) we will have:

$$\begin{cases} U_{22} = 0 \\ U_2 = U_{21} \end{cases}$$

   This means that the reflected wave is zero and there is no reflection; therefore, the voltage at the end of the line is the same as entering wave.

b. Open circuit line

   In this condition, the end of the line is open $(Z \to \infty)$. Therefore, the reflection coefficient will be 1 and the break coefficient will be 2. Therefore, using Equations (2.79) and (2.80), we will have:

$$\begin{cases} U_{22} = U_{21} \\ U_2 = 2U_{21} \end{cases}$$

   This means that the entering wave will completely be reflected.

c. Short circuit line

   In this case, the impedance at the end of line is equal to zero. The break coefficient is zero and the reflection coefficient is -1. Using Equations (2.79) and (2.80), we have:

$$\begin{cases} U_{22} = -U_{21} \\ U_2 = 0 \end{cases}$$

   This means that the entering wave will be reflected by a negative sign. In order to explain the matters in a more understandable way, we can use the example of a rope. Consider a rope, in which one side is connected to the wall and the other side is in our hand. When we move the rope once in an up position and the other in a down position, a semi-sine wave starts traveling along the x-axis and will be decayed during the travel. When the traveling wave reaches the dead-end wall, a portion of its energy moves to the wall and a portion of it gets reflected back to the rope. This example is the simulation of a real-life behavior of a traveling wave.

2.2.3 REFLECTION AT THE BEGINNING OF THE LINE

In order to analyze the reflection of a traveling wave at the beginning of the line, consider Figure 2.7. As mentioned before, the line has the characteristic impedance of Z_0.

Assuming at the sending end of the line, there is a source with the voltage U_0 and the internal impedance of Z_i. Impedance Z_2 is the terminating impedance at the end of the line. The traveling wave U_{11} is the wave moving toward the right side of the x-axis. Considering equation $I_{11} = \dfrac{U_{11}}{Z_0}$ and writing KVL, we will have Equations (2.83) and (2.84)

$$U_o = Z_i I_{11} + U_{11} = Z_i \frac{U_{11}}{Z_o} + U_{11} = U_{11}\left(\frac{Z_i}{Z_o} + 1\right) \tag{2.83}$$

$$U_{11} = U_o \frac{Z_o}{Z_i + Z_o} \tag{2.84}$$

The written KVL is for the time when the wave enters the line. As the time goes, the wave will travel through the line and will not stay stationary at the beginning of the line. At the starting point, the U_{11} travels along the line, and in this process, the amplitude of the wave will get attenuated and the amplitude U_{21} reaches the end of the line. This wave will get reflected encountering the end of the line. The reflected wave at the end of the line is called U_{22} and can be shown in Equation (2.85).

$$U_{22} = U_{21} \frac{Z_2 - Z_o}{Z_2 + Z_o} \tag{2.85}$$

The amplitude of U_2 can be found in Equation (2.86)

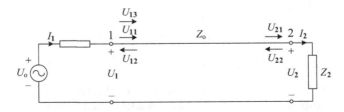

FIGURE 2.7 Transmission line with source impedance.

$$U_2 = U_{21} \frac{2Z_2}{Z_2 + Z_o} \qquad (2.86)$$

The wave U_{22} is traveling toward the sending end of the line and will be attenuated to the new magnitude of U_{21}. When this wave reaches node 1, some of it will be reflected and will move toward the end of the line to reach node 2. We call this wave U_{13}. Voltage at each point and each time spot is the addition of entering and reflecting waves. The same statement is true for current waves. Voltage and current waveforms are bounded with the characteristic impedance of the line. Therefore, the current at each point and each instant is the sum of all the waves at that point at a specific instant. At the instant t, during which the wave has traveled more than two times the length of the line, the voltage, and current at node 1 can be written using Figure 2.7, as follows:

$$\begin{cases} U_1 = U_{11} + U_{12} + U_{13} \\ I_1 = I_{11} + I_{12} + I_{13} \end{cases} \qquad (2.87)$$

Considering the direction illustrated for I_1 in Figure 2.7, we can write $I_1 = \dfrac{U_0 - U_1}{Z_i}$.

Using the mentioned equation for current, we will have:

$$I_{11} + I_{12} + I_{13} = \frac{U_o - U_{11} - U_{12} - U_{13}}{Z_i} \qquad (2.88)$$

Using Equations (2.83) and (2.88), we will obtain Equation (2.89)

$$(I_{11} + I_{12} + I_{13})Z_i = U_{11} \frac{Z_i + Z_o}{Z_o} - U_{11} - U_{12} - U_{13} \qquad (2.89)$$

Considering $U_{11} = Z_0 I_{11}$, $U_{12} = -Z_0 I_{12}$, $U_{13} = Z_0 I_{13}$ and Equation (2.51), we will obtain Equation (2.90)

$$U_{13} = U_{12} \frac{Z_i - Z_o}{Z_i + Z_o} \qquad (2.90)$$

Equation (2.90) implies that the wave U_{21} propagates from the end of the line toward the sending end of the line, and when it reaches the sending end, the wave gets reflected, only seeing the internal impedance of the source, that is, the wave will not see the source itself. For the wave that reaches point 1, it really does not matter if the source is providing a sine wave, a step function, or any other functions and what is important is the internal impedance of the source. The reason behind this lies in the high speed of the wave propagation, and therefore, the speed of the source variations in time will not be seen by the wave. This implies that the source can be considered a short circuit.

2.2.4 Voltage Distribution across a Lossless Line

In case of a lossless line, considering Equation (2.59), the α in equation $\gamma = \alpha + j\beta$ will be zero. Therefore, we will have:

$$\gamma = j\beta = j\omega\sqrt{L'C'} \tag{2.91}$$

$$v = \frac{\omega}{\beta} = \frac{1}{\sqrt{L'C'}} \tag{2.92}$$

$$Z_o = \sqrt{\frac{L'}{C'}} \tag{2.93}$$

We can conclude from the above equations that the speed and the characteristic impedance of the line are constant values. Let us remember the following general equation:

$$u(x,t) = Re\left\{\overline{u}_1 e^{-\alpha x} e^{j(-\beta x + \omega t)} + \overline{u}_2 e^{+\alpha x} e^{j(\beta x + \omega t)}\right\} \tag{2.94}$$

At $x = 0$, we will have:

$$u(0,t) = Re\left\{\overline{u}_1 e^{j\omega t} + \overline{u}_2 e^{j\omega t}\right\} = Re\left\{(\overline{u}_1 + \overline{u}_2)e^{j\omega t}\right\} \tag{2.95}$$

Let us assume that the voltage distribution at the beginning of the line is a sine wave and can be written as Equation (2.96)

$$u(0,t) = \overline{u}_o \cos(\omega t) = Re\left\{\overline{u}_o e^{j\omega t}\right\} \tag{2.96}$$

Using Equations (2.95) and (2.96), we will have:

$$\overline{u}_1 + \overline{u}_2 = \overline{u}_o \tag{2.97}$$

For the sake of simplicity, we assume that the derivative of the voltage at the beginning of the line is zero. Therefore:

$$\frac{\partial u(x,t)}{\partial x}\bigg|_{x=o} = 0 \Rightarrow Re\left\{\overline{u}_1(-\beta)e^{j\omega t} + \overline{u}_2\beta e^{j\omega t}\right\} = 0 \tag{2.98}$$

Therefore:

$$\overline{u}_1 = \overline{u}_2 \tag{2.99}$$

Considering Equation (2.97), we will have:

$$\overline{u}_1 = \overline{u}_2 = \frac{1}{2}\overline{u}_o \tag{2.100}$$

Applying the recent equation in (2.94) for a lossless line ($\alpha = 0$),

$$u(x,t) = Re\left\{\frac{\bar{u}_o}{2}e^{j(-\beta x+\omega t)} + \frac{\bar{u}_o}{2}e^{j(\beta x+\omega t)}\right\} = Re\left\{\bar{u}_o e^{j\omega t}\left(\frac{e^{-j\beta x}+e^{-j\beta x}}{2}\right)\right\} \quad (2.101)$$

Or

$$u(x,t) = Re\{\bar{u}_o e^{j\omega t}\cos(\beta x)\} \quad (2.102)$$

Therefore:

$$u(x,t) = \bar{u}_o\cos(\omega t)\cos(\beta x) \quad (2.103)$$

This waveform has been plotted in Figure 2.8 for different time spots.

Along the line, for any position that $\dfrac{\partial u}{\partial x} = 0$ at any given time, the current at that point has to be zero also. That can be concluded from Equation (2.104).

$$-\frac{\partial u}{\partial x} = R'i + L'\frac{\partial i}{\partial t} \quad (2.104)$$

We can write Equation (2.105) at the end of the line which is open.

$$\left.\begin{array}{l}-\dfrac{\partial u}{\partial x} = R'i + L'\dfrac{\partial i}{\partial t} \\[2mm] i(x)\big|_{x=l_{end}} = 0\end{array}\right\} \Rightarrow -\frac{\partial u}{\partial x} = 0 \quad (2.105)$$

Meaning at the open circuit line, $\dfrac{\partial u}{\partial x}$ is always zero.

Figure 2.9 shows voltage waves at different time stamps. Here, the length of the line is half of the length of the line illustrated in Figure 2.8.

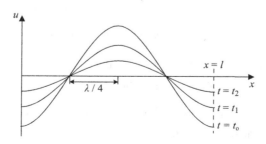

FIGURE 2.8 Voltage variations at different times and positions in a lossless line (sine wave input and open-ended line).

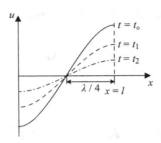

FIGURE 2.9 Voltage variations at different times and positions in a lossless shorter line (sine wave input and open-ended).

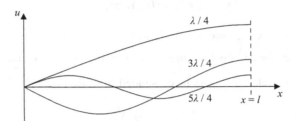

FIGURE 2.10 Voltage along the open circuit line.

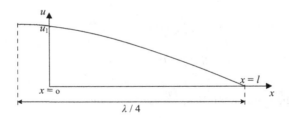

FIGURE 2.11 Voltage distribution along a short circuit line with the length $\frac{\lambda}{4}$.

For open circuit lines with lengths of $\frac{\lambda}{4}, \frac{3\lambda}{4}, \frac{5\lambda}{4}$ and so on, the voltage at the end of the line becomes larger than the voltage at the beginning of the line. Figure 2.10 shows this fact.

Obviously, if the end of the line is short circuited, we will have zero voltage at the end of the line. Figure 2.11 illustrates this fact. If the line has the length of $\frac{\lambda}{4}$, distribution of the voltage along the line will look like a cosine wave. In this case, voltage at the beginning of the line is u_1, and at $x = \frac{\lambda}{4}$, the voltage is zero (short circuit line). In case of having a line with length longer than $\frac{\lambda}{4}$, it is possible to have a voltage

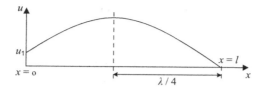

FIGURE 2.12 Voltage distribution along a short circuit line with the length $\frac{\lambda}{2}$.

higher than the source voltage somewhere along the line, while the end of the line is short circuited. This fact is illustrated in Figure 2.12.

2.2.5 FREQUENCY-DEPENDENT WAVE SPEED PROPAGATION VARIATION

a. Lossless line

In order to investigate the speed of the wave propagation along a transmission line, let us consider an overhead line. The propagation speed for a lossless line with known parameters for inductance and capacitance as L' and C' can obtained using Equation (2.106).

$$v = \frac{1}{\sqrt{L'C'}} \tag{2.106}$$

In order to find the parameters L' and C', we will use Equation (2.107). This equation will provide the average length of the line to ground. Figure 2.13 shows an overhead line installed between the two towers. As illustrated in the figure, T is the height of the line at the beginning of the line span and M is the height of the line in the middle of the span.

$$h = \frac{\sqrt{T^2 - M^2}}{\ln\left[\dfrac{T}{M} - \sqrt{\left(\dfrac{T}{M} \right)^2 - 1} \right]} \tag{2.107}$$

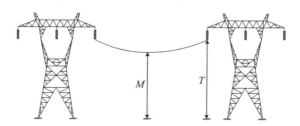

FIGURE 2.13 Profile of a line between two transmission lines.

Utilizing Taylor expansion equation on (2.107), we will obtain Equation (2.108)

$$h \cong M + \frac{1}{3}(T - M) \tag{2.108}$$

Now that we know the average length, we can use the famous Equations (2.109) and (2.110) to find L' and C'.

$$L' = 2 \times 10^{-7} \ln\left(\frac{2h}{r}\right) \quad \left[\frac{H}{m}\right] \tag{2.109}$$

$$C' = \frac{\varepsilon_r \times 10^{-9}}{18 \ln\left(\frac{2h}{r}\right)} \quad \left[\frac{F}{m}\right] \tag{2.110}$$

Implementing the last two equations into (2.106), we will have:

$$v = \frac{1}{\sqrt{\varepsilon_o \varepsilon_r \mu_o \mu_r}} \tag{2.111}$$

For a bare conductor parameter, μ_r and ε_r are 1. Knowing:

$$\mu_o = 4\pi \times 10^{-7} \quad \frac{N.s^2}{C^2}$$

$$\varepsilon_o = 8/85 \times 10^{-12} \quad \frac{C^2}{N.m^2}$$

Therefore, the speed will be calculated:

$$v = 3 \times 10^8 \quad \frac{m}{s}$$

Therefore, in case of a lossless line, the speed of the wave is the same as speed of the light. In Figure 2.14, the passing current through the inductor charges it and we call the energy stored in the inductor (W_L). Next, the energy

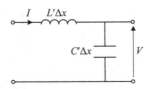

FIGURE 2.14   Lossless line with the length of Δx.

stored in the inductor will be discharged into the capacitor and become the electric field energy of the capacitor that we call (W_c); therefore, the wave will have a delay in moving.

The capacitor energy will be stored in the electric field and the inductor energy will be stored in the magnetic field as written in Equations (2.112) and (2.113).

$$W_C = \frac{1}{2} C' \Delta x V^2 \tag{2.112}$$

$$W_L = \frac{1}{2} L' \Delta x I^2 \tag{2.113}$$

In the lossless condition, the amount of energy stored in the magnetic field is exactly the same as the amount of energy stored in the electric field. This fact will give Equations (2.114) and (2.115)

$$C'V^2 = L'I^2 \tag{2.114}$$

$$\frac{V}{I} = \sqrt{\frac{L'}{C'}} = Z_0 \tag{2.115}$$

As we knew, the relationship between the voltage and current will be established through the characteristic impedance. This impedance is a fixed value and is independent of frequency.

b. Lossy line

We follow this topic using Equations (2.59) and (2.60) repeated here as Equations (2.116) and (2.117). Parameters α and β are plotted as a function of frequency in Figure 2.15. The frequency domain has been divided into three different domains as low frequency, mid frequency, and high frequency.

$$\alpha = \sqrt{\frac{1}{2}(R'G' - \omega^2 L'C') + \frac{1}{2}\sqrt{(R'^2 + \omega^2 L'^2)(G'^2 + j\omega^2 C'^2)}} \tag{2.116}$$

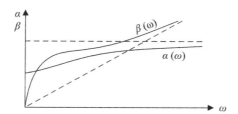

FIGURE 2.15   Parameters α and β with frequency.

$$\beta = \sqrt{\frac{1}{2}(\omega^2 L'C' - R'G') + \frac{1}{2}\sqrt{(R'^2 + \omega^2 L'^2)(G'^2 + j\omega^2 C'^2)}} \qquad (2.117)$$

In the low-frequency domain, $\omega'L' \ll R'$ and $\omega'C' \ll G'$. Implementing the mentioned conditions in Equations (2.59) and (2.60), we will have:

$$\alpha \cong \sqrt{R'G'} \qquad (2.118)$$

$$\beta = 0 \qquad (2.119)$$

Therefore,

$$v = \frac{\omega}{\beta} = \frac{2\sqrt{R'G'}}{R'C' + L'G'} \qquad (2.120)$$

The condition mentioned above is still valid for mid-range frequency and $\omega'L' \ll R'$ is still a valid condition. But the situation changes for the capacitor and we will have $\omega'C' \gg G'$. Considering the abovementioned conditions in Equations (2.59) and (2.60), we will have:

$$\gamma = \sqrt{j\omega R'C'} = (1+j)\sqrt{\frac{1}{2}\omega R'C'} \qquad (2.122)$$

$$\alpha = \beta = \sqrt{\frac{1}{2}\omega R'C'}$$

Therefore,

$$v = \frac{\omega}{\beta} = \sqrt{\frac{\omega}{\frac{1}{2}R'C'}} \qquad (2.123)$$

In the high-frequency domain, we will have $\omega'L' \gg R'$ and $\omega'C' \gg G'$. Therefore,

$$\beta = \omega\sqrt{L'C'} \qquad (2.124)$$

$$\alpha = \frac{R'}{2}\sqrt{\frac{C'}{L'}} + \frac{G'}{2}\sqrt{\frac{L'}{C'}} \qquad (2.125)$$

Therefore,

$$v = \frac{\omega}{\beta} = \frac{1}{\sqrt{L'C'}} \qquad (2.126)$$

In the situation, known as distortionless, the following equation is true:

$$\frac{R'}{L'} = \frac{G'}{C'} \tag{2.127}$$

Considering Equation (2.172), we will have:

$$\alpha = R'\sqrt{\frac{C'}{L'}} \tag{2.128}$$

$$\beta = \omega\sqrt{L'C'} \tag{2.129}$$

$$v = \frac{1}{\sqrt{L'C'}} \tag{2.130}$$

Transient waveforms can be expanded using Fourier series: therefore, the transient waveform will be presented as a series of sinusoidal waves with different frequencies. Considering Equation (2.117) and the fact that β changes with frequency, it will be concluded that there will be a speed difference for different sinusoidal parts of the transient wave, and this will cause different propagation speeds for different parts or scattering of different parts of the wave as they propagate.

The parameter α is the attenuation parameter. This parameter has different values for different frequencies. Therefore, the wave components will be attenuated with different values. In summary, different values for α and β at different frequencies will impact the speed and attenuation of the wave at different frequencies; therefore, we will have scattering phenomenon and distortion of the waveform as the wave propagates. Figure 2.15 shows the trends that these parameters are going through as the frequency changes.

2.2.6 FREQUENCY-DEPENDENT CHARACTERISTIC IMPEDANCE VARIATIONS

Transmission line characteristic impedance can be obtained using Equation (2.131)

$$Z_o = \sqrt{\frac{z'}{y'}} \tag{2.131}$$

Knowing $z' = R' + j\omega L'$ and $y' = G' + j\omega C'$ and implementing them in Equation (2.132), we will have:

$$Z_o = \sqrt{\frac{R' + j\omega L'}{G' + j\omega C'}} = Z_1 + jZ_2 \tag{2.132}$$

In the above equation, Z_1 is the real part of the characteristic impedance and Z_2 is the imaginary part of it. The sign for Z_2 will be obtained from $G'L' - R'C'$.

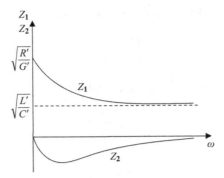

FIGURE 2.16 Z_1 and Z_2 versus frequency.

Figure 2.16 shows the parameters Z_1 and Z_2 and their change as the frequency changes. As it can be seen in that figure, the imaginary part of the impedance is zero when the frequency is zero and eventually becomes zero again as the frequency increases. At higher frequencies, the limit of the characteristic impedance will tend to $\sqrt{\dfrac{L'}{C'}}$.

2.3 STEP-FORM TRAVELING WAVES

So far, we have studied the traveling waves for the sinusoidal waves. In the case of a step wave, we can expand the wave using Fourier transformation and look at the problem as a combination of sine waves. Therefore, the study will be very similar to what has been discussed in Section 2.2. We are not going to examine what is explained here, but rather we consider the step function as a function $f(t)$ and will apply it to the beginning of the line, and we will find the voltage wave propagating through the line. The reasons for considering step function in transient studies are summarized as follows:

1. The speed of the wave propagation is so high in comparison with the variation of sine wave of the generation system at power frequency that during transient studies, it can be considered as a function with a constant value, that is, the step function.
2. Transient conditions usually happen due to lightning and switching, and therefore, the wave that appeared due to these interactions will be an impulse wave. Considering the impulse function for transient studies will complicate the analysis. In this case, we should consider the wave rise time and fall-time. These parameters vary with cases and therefore influence the study results. The rise time of a step function is zero and its fall-time is infinite. This will help easing the calculations in transient analysis.
3. Impulse function will reach to its highest value in a very short amount of time, and in that time, its magnitude decays. The step function can cause more stress in the system compared with the impulse function. Replacing

the study of the impulse function with the step function, we are considering the worst-case scenario, as the step function reaches its highest value in zero microsecond and stays at its highest amplitude without any decay. Therefore, the design under these conditions should be valid for real conditions.

2.3.1 SOLVING EQUATIONS

Considering Equations (2.11) and (2.15), we will have:

$$\frac{\partial^2 u}{\partial x^2} = R'G'u + (R'C' + L'G')\frac{\partial u}{\partial t} + L'C'\frac{\partial^2 u}{\partial t^2} \tag{2.133}$$

$$\frac{\partial^2 i}{\partial x^2} = R'G'i + (R'C' + L'G')\frac{\partial i}{\partial t} + L'C'\frac{\partial^2 i}{\partial t^2} \tag{2.134}$$

Solving the abovementioned equations for an arbitrary function of $f(t)$ will be hard and needs numerical computation methods. Let us assume that R' and G' are zero, that is, we are considering the lossless condition.

$$\begin{cases} \dfrac{\partial^2 u}{\partial x^2} = L'C'\dfrac{\partial^2 u}{\partial t^2} \\ \dfrac{\partial^2 i}{\partial x^2} = L'C'\dfrac{\partial^2 i}{\partial t^2} \end{cases} \tag{2.135}$$

$$\begin{cases} \dfrac{\partial^2 u}{\partial x^2} = \dfrac{1}{v^2}\dfrac{\partial^2 u}{\partial t^2} \\ \dfrac{\partial^2 i}{\partial x^2} = \dfrac{1}{v^2}\dfrac{\partial^2 i}{\partial t^2} \end{cases} \tag{2.136}$$

The abovementioned equations are one-dimensional equations. In conditions such as propagation of electric wave in transmission line or resonating wave in a rope or wire, we have the same condition. Solution for the abovementioned equation is a function of time and position, that is:

$$u = u(x,t) \tag{2.137}$$

Knowing the initial and boundary conditions of differential equations will give a unique solution to them. The boundary conditions determine the current and voltage at the sending and receiving ends of the line, and the initial conditions are the current, voltage, and their derivative values with respect to t across the line at $t = 0$. Assume that we have:

$$u(x,t)\big|_{t=0} = f(x) \tag{2.138}$$

$$\left.\frac{\partial u(x,t)}{\partial t}\right|_{t=0} = 0 \tag{2.139}$$

Let us consider the method of separation of variables for solving the equations. This means that we have assumed to have a solution in the following form:

$$u(x,t) = F(x)H(t) \tag{2.140}$$

This equation divides the $u(x, t)$ into two different functions, one is completely function of time and independent of x and the other is a function of x and totally independent of time. Considering Equation (2.136), we will have:

$$\frac{d^2H}{dt^2}F = v^2H\frac{d^2F}{dx^2} \tag{2.141a}$$

$$\frac{\frac{d^2H}{dt^2}}{v^2H} = \frac{\frac{d^2F}{dx^2}}{F} \tag{2.141b}$$

As it can be seen that the left-hand side of the latest equation is a function of time and the right-hand side is the function of x, therefore, in order for the equation to make sense, both sides need to have constant values. This has been shown in Equation (2.141c)

$$\frac{\ddot{H}}{v^2H} = \frac{F''}{F} = k \tag{2.141c}$$

F'' is the second derivative of the function F with respect to position and $\ddot{H}$ is the second derivative of the function H with respect to time. Therefore, we have:

$$\frac{\ddot{H}}{v^2H} = k \tag{2.141d}$$

$$\frac{F''}{F} = k \tag{2.141e}$$

From the latest equation and assuming k a positive parameter, we have:

$$k = \alpha^2 \geq 0 \tag{2.141f}$$

$$F'' = kF \tag{2.141g}$$

This will be fairly easy to solve. Let us use the characteristic equation of differential equation:

$$m^2 - k = 0 \Rightarrow m^2 - \alpha^2 = 0 \Rightarrow m = \pm\alpha \tag{2.141h}$$

$$F(x) = Ae^{\alpha x} + Be^{-bx} \tag{2.142}$$

If we assume that the voltage u is limited at each point of the line, then α needs to be a complex number. Otherwise, with the increase in x, the value of F will be increased exponentially. Obviously, in order for the α to be a complex number, K needs to be negative.

$$k = -q^2 \tag{2.143}$$

Therefore, the differential equation will be:

$$F'' + q^2 F = 0 \tag{2.144}$$

The solution for this equation will be

$$F(x) = a \cos qx + b \sin qx \tag{2.145}$$

Also, we have:

$$u(x,t)\big|_{t=0} = f(x) \tag{2.146}$$

Using the Fourier series, we can expand the $f(x)$ as in Equation (2.147).

$$f(x) = \sum_{n=0}^{\infty} a_n \cos \frac{n\pi}{l} x + b \sum_{n=0}^{\infty} a_n \sin \frac{n\pi}{l} x$$
$$= a_o + \sum_{n=1}^{\infty} a_n \cos \frac{n\pi}{l} x + b \sum_{n=1}^{\infty} a_n \sin \frac{n\pi}{l} x \tag{2.147}$$

Here, l is the length of the line. This equation is the solution for the differential Equation (2.145). Considering the fact that $H(t)$ at $t = 0$ has to be 1. From Equation (2.147), we can conclude that k is negative and we have Equation (2.148)

$$k = -q^2 = -\left(\frac{n\pi}{l}\right)^2 \tag{2.148}$$

From Equation (2.141d) and using the specified value for k, we can find:

$$\ddot{H} + \left(\frac{n\pi}{l} v\right)^2 H = 0 \tag{2.149}$$

$$\ddot{H} + \lambda_n^2 H = 0 \tag{2.150}$$

$$\lambda_n \overset{\Delta}{=} \frac{n\pi}{l} v \tag{2.151}$$

In Equation (2.151), for each n, we will have:

$$H_n(t) = D_n \cos \lambda_n t + E_n \sin \lambda_n t \qquad (2.152)$$

Therefore, the general solution for the differential equation will be as:

$$
\begin{aligned}
u(x,t) &= \sum_{n=0}^{\infty} u_n(x,t) \\
&= \sum_{n=0}^{\infty} \left[\left(a_n \cos \frac{n\pi}{l} x + b_n \sin \frac{n\pi}{l} x \right) (D_n \cos \lambda_n t + E_n \sin \lambda_n t) \right]
\end{aligned}
\qquad (2.153)
$$

Figure 2.17 shows the last equation at different time stamps as a function of position.

For setting $n = 0$, Equation (2.153) will bring a constant value that is obtained from a_0 in Equation (2.147). Equation (2.153) is called the stationary wave equation. This equation can be justified as Fourier expansion of the initial condition $u(x, t = 0)$. Therefore, the equation will be presented as a sum of sine and cosine functions. In this equation, each sine or cosine function is fixed at its particular position (i.e. no change in position) and oscillates with the frequency of $f_n = \dfrac{\lambda_n}{2\pi}$. For each n, we have a time sine equation and a position sine equation. Considering Equation (2.151), that is, $\lambda_n = \dfrac{n\pi}{l} v$, the oscillation frequency of the nth piece can be written as follows:

$$f_n = \frac{\lambda_n}{2\pi} = \frac{n\pi}{l} \frac{v}{2\pi} = \frac{nv}{2l} \qquad (2.154)$$

As Equation (2.153) illustrates, the voltage magnitude for $u(x, t)$ is the sum of all the stationary waves. Therefore, the main wave is not stationary and travels along the line. Consider the fact that this wave has stationary components. For each n in this equation, an Eigen function will be obtained. In this case, λ_n is the Eigen value of the equations.

Using Equation (2.153) with a little change, the traveling wave equation will be obtained. We assume at $t = 0$ the voltage along the line is $u_0(x)$ and the derivative of

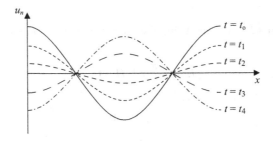

FIGURE 2.17   Stationary sine wave for $n = 2$.

that, that is, $\dfrac{\partial u}{\partial t}$ is zero in all places. Having the abovementioned initial conditions, we can solve the differential equation. Applying $\dfrac{\partial u}{\partial t} = 0$ in (2.153) will give:

$$E_n = 0 \qquad (2.155a)$$

Using the first initial condition, we will have $D_n = 1$.
Therefore,

$$u(x,t) = \sum_{n=0}^{\infty} \left(a_n \cos \frac{n\pi}{l} x + b_n \sin \frac{n\pi}{l} x \right) \cos \lambda_n t \qquad (2.155b)$$

$$u(x,t) = \sum_{n=0}^{\infty} a_n \cos \frac{n\pi}{l} x \cos \lambda_n t + \sum_{n=0}^{\infty} b_n \sin \frac{n\pi}{l} x \cos \lambda_n t \qquad (2.155c)$$

Consider the following trigonometric identities,

$$\cos x \cos y = \frac{1}{2}[\cos(x+y) + \cos(x-y)] \qquad (2.156)$$

$$\sin x \cos y = \frac{1}{2}[\sin(x+y) + \sin(x-y)] \qquad (2.157)$$

we have:

$$u(x,t) = \frac{1}{2} \sum_{n=0}^{\infty} a_n \left[\cos\left(\frac{n\pi}{l} x + \lambda_n t \right) + \cos\left(\frac{n\pi}{l} x - \lambda_n t \right) \right]$$

$$+ \frac{1}{2} \sum_{n=0}^{\infty} b_n \left[\sin\left(\frac{n\pi}{l} x + \lambda_n t \right) + \sin\left(\frac{n\pi}{l} x - \lambda_n t \right) \right] \qquad (2.158)$$

With $\lambda_n = \dfrac{n\pi}{l} v$, we can write:

$$u(x,t) = \frac{1}{2} \sum_{n=0}^{\infty} a_n \left[\cos \frac{n\pi}{l} (x+vt) + \cos \frac{n\pi}{l} (x-vt) \right]$$

$$+ \frac{1}{2} \sum_{n=0}^{\infty} b_n \left[\sin \frac{n\pi}{l} (x+vt) + \sin \frac{n\pi}{l} (x-vt) \right] \qquad (2.159)$$

It is possible to write this equation in a simple form of:

$$u(x,t) = \frac{1}{2}[f(x+vt)+f(x-vt)] \qquad (2.160a)$$

Here,

$$f(x+vt) = \sum_{n=0}^{\infty} a_n \cos\frac{n\pi}{l}(x+vt) + b_n \sin\frac{n\pi}{l}(x+vt) \qquad (2.160b)$$

$$f(x-vt) = \sum_{n=0}^{\infty} a_n \cos\frac{n\pi}{l}(x-vt) + b_n \sin\frac{n\pi}{l}(x-vt) \qquad (2.160c)$$

We see a traveling wave in Equation (2.160a), as t changes with Δt and x with Δx, where $\Delta x = v\Delta t$, therefore, $x - vt$ will yield a constant value, and we have

$$[x+\Delta x - v(t+\Delta t)] = [x+(\Delta x - v\Delta t) - vt] = (x-vt) \qquad (2.161)$$

If t changes with Δt and x with Δx, where $\Delta x = v\Delta t$, therefore, $x + vt$ will yield a constant value, and we have

$$[x+\Delta x + v(t+\Delta t)] = [x+(\Delta x + v\Delta t) + vt] = (x+vt) \qquad (2.162)$$

Equations (2.161) and (2.162) will tell us that we have a traveling wave with the speed of v propagation on the right and left side of the x-axis, that is, the positive x and negative x. Figure 2.18 shows these waves.

Let us remind ourselves of the initial conditions one more time. At $t = 0$, voltage is $u_0(x)$ and the derivative of that meaning $\frac{\partial u}{\partial t} = 0$.

$$u(x,t=0) = u_o(x) \qquad (2.163)$$

Therefore, from Equation (2.160a), we will have:

$$u(x,t) = \frac{1}{2}[u_o(x+vt)+u_o(x-vt)] \qquad (2.164a)$$

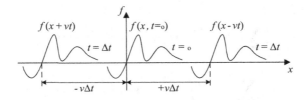

FIGURE 2.18 Traveling waves toward positive and negative x.

This equation tells us that the initial condition of the voltage is divided into two equal parts, one of them, as a traveling wave, will propagate with the speed of v toward positive x and the other one with the same speed will move toward negative x. In order for these two waves to have same the amplitude, $\frac{\partial u}{\partial t} = 0$ has to be zero at $t = 0$, that is:

$$\left.\frac{\partial u}{\partial t}\right|_{t=0} = 0 \qquad\qquad (2.164\text{b})$$

2.3.2 CURRENT EQUATIONS (D'ALEMBERT'S SOLUTION)

Using the D'Alembert's solution methodology for partial differential equations (PDE), we can directly obtain Equation (2.160a). In this approach, the solution of $u(x, t)$ will be written as follows:

$$u(x,t) = \Phi(x + ct) + \Psi(x - ct) \qquad\qquad (2.165)$$

We will discuss how to obtain values for c.

We can easily show that this is the solution for the single-dimension wave differential equation, that is, Equation (2.135), which we repeat in here:

$$\frac{\partial^2 u}{\partial x^2} = L'C' \frac{\partial^2 u}{\partial t^2} \qquad\qquad (2.166)$$

Let us take double derivatives with respect to time and position from Equation (2.165).

$$\frac{\partial^2 u}{\partial x^2} = \frac{\partial^2 \Phi}{\partial x^2} + \frac{\partial^2 \Psi}{\partial x^2} \qquad\qquad (2.167)$$

$$\frac{\partial^2 u}{\partial t^2} = c^2 \frac{\partial^2 \Phi}{\partial t^2} + c^2 \frac{\partial^2 \Psi}{\partial t^2} \qquad\qquad (2.168)$$

Implementing the derivatives into the single dimension wave differential equation, we will have $c^2 = \frac{1}{L'C'}$. Again, Equation (2.165) shows a traveling wave.

Figure 2.19 shows the line voltage at $t = 0$ in the form of $u_0(x)$. Assuming $\frac{\partial u}{\partial t} = 0$ at $t = 0$, the initial voltage $u_0(x)$ is being divided into two equal portions and each portion travels with the same speed along the x-axis but in opposite directions.

We can have the same solution for current with some changes in the initial conditions. For voltage at $t = 0$, we have the initial conditions, $u_0(x)$ and $\frac{\partial u}{\partial t} = 0$.

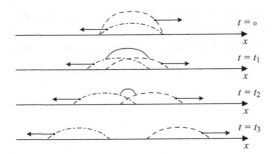

FIGURE 2.19 Voltage at different time stamps, assuming $\left.\dfrac{\partial u}{\partial t}\right|_{t=0}=0$.

Considering these initial conditions and the equation $-\dfrac{\partial i}{\partial x}=G'u+C'\dfrac{\partial u}{\partial t}$ and ignoring G' we will have:

$$i(x,t)\,\big|_{t=0}=0 \tag{2.169}$$

We ignore the constant of integration and that will not change the integrity of the wave. Considering Equation (2.111) and the initial condition for the current, we will have:

$$i(x,t)=\frac{1}{2}[h(x+vt)-h(x-vt)] \tag{2.170}$$

This equation at $t=0$ and before any movements of the electrons is zero.

2.3.3 FORWARD AND BACKWARD WAVES

According to Equation (2.22) and in general condition, we have:

$$u(x,t)=\underbrace{f_1(t)e^{\gamma x}}_{u_r}+\underbrace{f_2(t)e^{-\gamma x}}_{u_f} \tag{2.171}$$

In the above equation, f_1 moves to the negative side of the x-axis and f_2 moves to the positive side of the x-axis. The wave traveling to the right-hand side is called the forward wave and it is shown to be the index f. The traveling wave moving toward the negative side of the x-axis is called the backward wave and it is shown with r index. According to Equation (2.24), we have:

$$\frac{\partial u}{\partial x}=-z'i \tag{2.172}$$

Therefore, we will have:

$$i = -\frac{\gamma}{z'}[f_1(t)e^{\gamma x} - f_2(t)e^{-\gamma x}]$$

(2.173)

Considering the fact that $\gamma = \sqrt{z'y'}$ and $Z_0 = \sqrt{\dfrac{z'}{y'}}$ we will have:

$$i = -\frac{1}{Z_o}[f_1(t)e^{\gamma x} - f_2(t)e^{-\gamma x}]$$

(2.174)

We define i_r and i_f as the following

$$i_r \overset{\Delta}{=} -\frac{1}{Z_o} f_1(t)e^{\gamma x}$$

(2.175)

$$i_f \overset{\Delta}{=} \frac{1}{Z_o} f_2(t)e^{-\gamma x}$$

(2.176)

And apparently,

$$i = i_f + i_r$$

(2.177)

$$\begin{cases} i_f = \dfrac{u_f}{Z_o} \\[2mm] i_r = -\dfrac{u_r}{Z_o} \end{cases}$$

(2.178)

The general equation for the forward and backward waves can be written as:

$$\begin{cases} u = u_f + u_r \\ i = i_f + i_r \\ u_f = Z_o i_f \\ u_r = -Z_o i_r \end{cases}$$

(2.179)

In Figure 2.20, the current and voltage waves are shown in a general form. The assumption is that there is voltage on the line at $t = 0$ and it will not change until $t = 0$; therefore, the current is zero until $t = 0$. This voltage may appear due to the capacitor inductions. For instance, when there is a charged cloud above the transmission line, some amount of electric charge with the opposite polarity will be stored on the line. When the cloud passes above the line or in case of a lightening, that is, electric discharge of the cloud, the electric charge of the line will get depleted. Figure 2.21 illustrates the accumulation of the electric charge and its travel in the opposite directions.

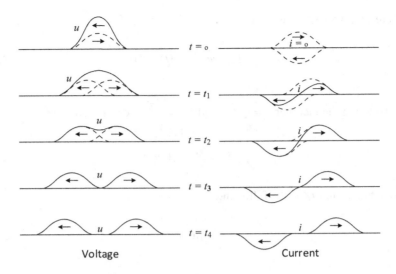

FIGURE 2.20 Voltage and current waves traveling to the positive and negative side of the *x*-axis.

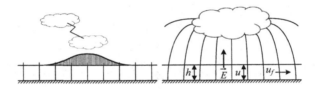

FIGURE 2.21 Charged cloud influence on appearance of the voltage on overhead line.

2.3.4 REFLECTION RULES FOR STEP WAVES

When an electromagnetic wave travels along a transmission line, after reaching the end of the line, a portion of it will get reflected. Reflection rules can be easily obtained using ohm' law, Kirchhoff's law, and what is illustrated in Equation (2.179). We analyze some cases of reflection.

a. Open Circuit Line
   The end of the line is open circuited. In this case, when the wave reaches the end of the line, according to Kirchhoff's law, the sum of the currents should be zero, and therefore:

$$i = i_f + i_r = 0 \tag{2.180}$$

$$i_f = -i_r \tag{2.181}$$

This means that there will be a line current of i_r. This current starts flowing in the opposite direction of the original wave. This current will cause a voltage along the line. Using $u_r = -Zi_r$, we have:

$$u_r = -Zi_r = -Z(-i_f) = Zi_f = u_f \tag{2.182}$$

Therefore,

$$u = u_f + u_r = 2u_f \tag{2.183}$$

The voltage wave will get reflected when it reaches the end of the line. The reflected wave is equal to the forward wave, and the voltage at the end of the line is the sum of forward and backward waves, that is, two times of the forward voltage wave. Current is zero at the positions where there are forward and backward waves. Figure 2.22 shows four different time spans of the wave. These different times are labeled as a, b, c, and d.

In part a of this figure, the wave has not reached the end of the line yet. In part b, the wave has reached the end of the line but has not been reflected yet, and in part c, the wave has reflected and has traveled through a small portion of line in the opposite direction. Finally, in part d, the reflected wave has traveled a larger piece of the line in the opposite direction.

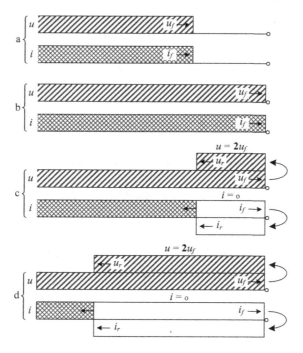

FIGURE 2.22   Wave reflection at the end of an open circuit line.

b. Short Circuit Line

In this case, the end of the line is short circuited. Therefore, voltage at the end of the line is zero.

$$u = u_f + u_r = 0 \qquad (2.184)$$

$$u_f = -u_r \qquad (2.185)$$

In this case, voltage u_r will appear that travels in the opposite direction of the line and makes current in the line. We can write:

$$i_r = -\frac{u_r}{Z_o} = -\frac{-u_f}{Z_o} = \frac{u_f}{Z_o} = i_f \qquad (2.186)$$

Therefore:

$$i = i_f + i_r = 2i_f \qquad (2.187)$$

The voltage wave will get reflected after reaching the end of the line. The reflected (backward) voltage wave is the same as the forward wave with an opposite sign. Along the line, wherever there is backward and forward voltage wave, the voltage is zero. For current, wherever there is backward and forward waves, it is twice the forward wave. Figure 2.23 shows four different time spans of this condition similar (dual) to Figure 2.23.

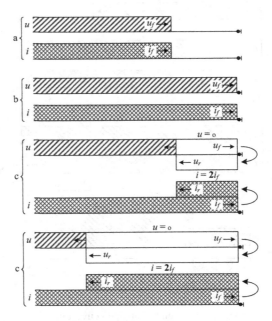

FIGURE 2.23 Wave reflection in a short circuited line.

c. Resistor at the End of the Line

In this condition, end of the line is terminated by the resistor R. In this situation, the relationship between voltage and current is:

$$u = Ri_R \qquad (2.188)$$

When the wave reaches the end of the line, we will have:

$$u = u_f + u_r = Z_o i_f - Z_o i_r \qquad (2.189)$$

$$i = i_f + i_r = i_R \qquad (2.190)$$

Therefore:

$$i_r = -i_f + i_R = -i_f + \frac{u}{R} \qquad (2.191)$$

$$u = Z_o i_f + Z_o i_f - u\frac{Z_o}{R} \qquad (2.192)$$

$$u\left(1 + \frac{z_0}{R}\right) = 2Z_o i_f = 2u_f \qquad (2.193)$$

$$u = \frac{2R}{R + Z_o} u_f \qquad (2.194)$$

$$u_r = \frac{R - Z_o}{R + Z_o} u_f \qquad (2.195)$$

Now, we can define the break and reflection coefficients for step wave as in Equations (2.196)

$$\begin{cases} b = \dfrac{2R}{R + Z_o} \\[2mm] r = \dfrac{R - Z_o}{R + Z_o} \end{cases} \qquad (2.196)$$

If the resistance R is greater than Z_0, the voltage magnitude at the end of the line ($|u|$) will be greater than the forward voltage ($|u_f|$), that is, u_r has the same sign as u_f. On the other hand, if R is smaller than Z_0, $|u|$ will be smaller than $|u_f|$; therefore, the sign for u_r is the opposite of u. In both cases, $|u_r|$ is smaller than $|u_f|$.

2.3.5 LOSS CONSIDERATION IN LINE

Considering the parameters R' and G' complicates the form of differential equations. Therefore, the solution will not be the simple form of $u = g(x \mp vt)$. In this case, the speed of the traveling wave will be different regarding the frequency. Also, the wave impedance of the line is a function of frequency. Apparently, due to the existence of losses, the wave will be attenuated by damping coefficients, which differ with frequency. In order to analyze the case, we use Fourier transformation. If we look at Equation (2.58) one more time, we will have unique values for v (speed of traveling wave), Z (line impedance), and α (damping factor) at different frequencies. In this case, after some time, not only the wave is attenuated but also will be deformed. With the existence of different speeds at different frequencies, the wave will be scattered; in other words, the components of the Fourier series will be scattered through the line. The wave will have a longer forehead. First, the forehead of the wave consists of higher frequencies and those are damped faster. Also, the speed of the higher frequency waves is higher than the speed of the lower frequency waves. Let us assume the simplest possible solution for this case as it is suggested by:

$$u(x,t) = [u(x,t)\big|_{t=0}]e^{-\alpha x} \tag{2.197}$$

We have made quite a number of simplifications by considering a constant damping wave and ignoring the fact that the parameter α changes with frequency. Equation (2.197) suggests that switching the line farther away from the source will cause less damage as the wave has the attenuation that increases as the length increases. There are other considerations that need to be made in here. In transients for very high voltages (100 kV and above), the value for G' is a function of voltage. Because of the very high voltage, the electric field intensity on the conductor will be increased, and therefore, the atmosphere around the line will be ionized. As the applied voltage gets bigger, the value for G' gets larger too. This statement clarifies that the damping factor will be a function of voltage as well. Ionization intensity is related to the polarity of the voltage wave. Therefore, G' will be the function of both magnitude and polarity of the voltage.

On the contrary, due to the ionization, air around the wire will act as a conductor. This means that the effective cross-section of the line is increased as well. The increase in the effective cross-sectional area of the conductor will impact the value of C'; this will cause an in increase in scattering effect on the wave. Also, as the ionization increases, corona effect will appear along the line. This effect will increase the losses in the line.

In order to analyze a distorted waveform, we can divide the wave into a number of step waves and use the methodology explained here and the superposition effect will give the accumulated result of the distorted wave.

2.4   PSCAD EXAMPLES

Following is a set of examples related to problems discussed in this chapter. These examples have been modeled in PSCAD. The model includes a long transmission line with the following parameters.

Line length: 370 km
Nominal voltage: 215 kV
Nominal frequency: 50 Hz
Resistance: 0.0996 Ω/km
Inductive reactance: 0.5143 Ω/km
Capacitive reactance: 0.12173 × 10⁶ Ω/km

a. Switching of a DC Voltage Source, Open Circuit Line
 a.1 Simulation of lossless line
  In this example, the line is considered to be lossless. The resistance specified in the beginning of this subsection has been set to zero. Bergeron line
  model is used throughout the PSCAD models. Figure 2.24 illustrates the
  PSCAD single-line diagram of an open circuited long transmission line
  exposed to a DC source.
   At $t = 0.05$ s, the circuit breaker closes the DC source to the line. E_a and
  E_b are the sending end and receiving end voltages, respectively. Figure
  2.25 presents the voltage waves of E_a and E_b in time domain.
   As it is illustrated and measured in Figure 2.25, after closing the breaker
  at $t = 0.05$ s, it takes 2.5 ms for voltage wave to reach the end of the line.
  At this point, the voltage at the end of the line reaches to twice the nominal
  value of the voltage.
   Using Equation (2.29) along with the system parameters specified
  above, it is possible to calculate the traveling wave speed of propagation
  and also the time that takes for the wave to reach end of the line. The aforementioned calculations are presented below:

$$L' = \frac{X_L}{\omega} = \frac{0.5143 \times 10^{-3}}{2\pi \times 50} = 1.6367 \times 10^{-6} \text{ H/m}$$

$$C' = \frac{1}{\omega X_C} = \frac{1}{2\pi \times 50 \times 0.12173 \times 10^9} = 2.6149 \times 10^{-11} \text{ F/m}$$

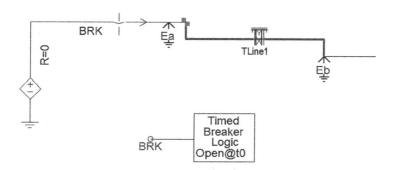

FIGURE 2.24　Single-line diagram of an open circuit test.

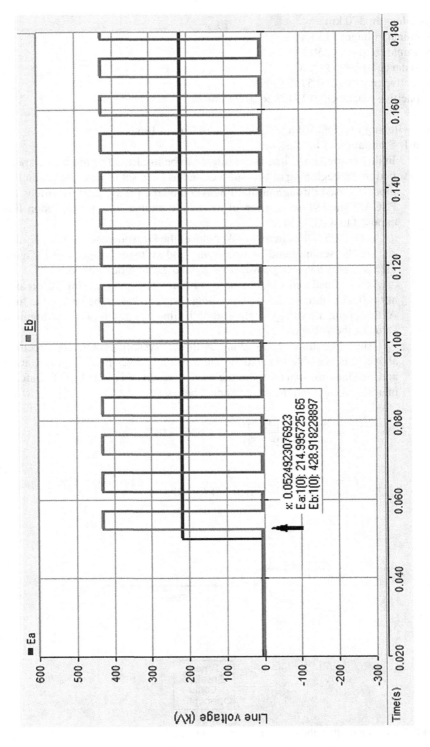

x: 0.052492307923
Ea:1[0]: 214.995725165
Eb:1[0]: 428.918228897

FIGURE 2.25 Sending end and receiving end voltages.

$$v = \frac{1}{\sqrt{L'C'}} = 152857909.6 \, \text{m/s}$$

$$T = \frac{\text{Line length}}{v} = \frac{370000 \, \text{m}}{152857909.6 \, \text{m/s}} = 2.42 \, \text{ms}$$

The L and C parameters of the line presented above are calculated in metric system (H/m and Fm). The above calculation provided the traveling time of 2.42 ms, which is very close to the observation made in the PSCAD plot.

The DC wave in this system does not get attenuated due to the fact that a lossless transmission line is studied.

a.2 Simulation of lossy line

The single-line diagram presented in Figure 2.24 has been considered in a lossy condition. Figure 2.26 presents the sending end and receiving end voltages of the line for this condition. The difference between the results in Figures 2.25 and 2.26 lies in the attenuation factor that can be seen in Figure 2.26. Resistance of the line attenuates the traveling wave, and after 0.15 s of switching, the sending end voltage reaches the magnitude of the receiving end voltage.

b. Switching of a DC Voltage Source, Short Circuit Line

b.1 Lossless line

Figure 2.27 presents a short circuited transmission line in PSCAD. This line is energized with a DC voltage source. The small resistance presented at the end of the line is a replacement for a short circuit that is suitable for PSCAD calculations. Circuit breaker closes the source to the line at $t = 0.05$ s. E_a and E_b are the sending end and receiving end voltages, respectively. Figure 2.28 illustrates these voltages as functions of time.

Although the breaker closes at $t = 0.05$ s, as it can be seen on Figure 2.28, voltage at the end of the line stays zero. As it is discussed in this chapter, the reflection coefficient is -1. Therefore, the wave that reaches end of the line gets reflected totally with a negative magnitude.

b.2 Lossy line

The same result as presented in Figure 2.28 is expected in here. The study has been carried out, and Figure 2.29, which is exactly the same as Figure 2.28, is presented for sending end and receiving end voltages.

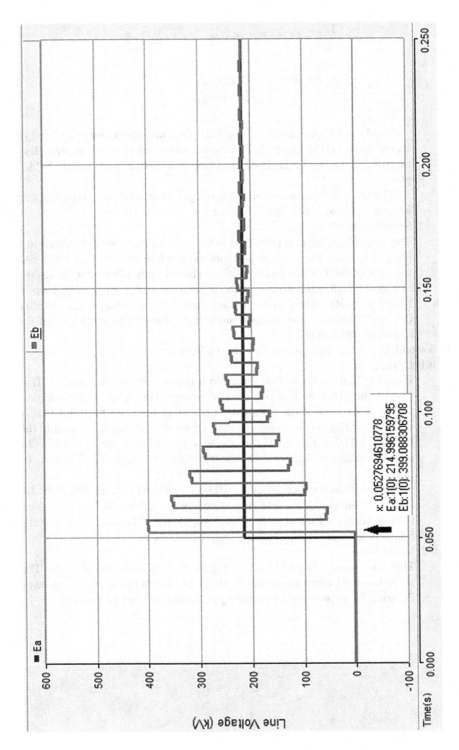

FIGURE 2.26 Voltages of the a.2 test.

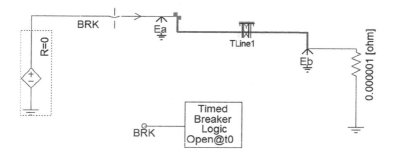

FIGURE 2.27   Single-line diagram of short circuit test.

c. Switching of a DC Voltage Source, Surge Impedance Load

Here, the same system as before is terminated to a resistance with the same value as the line characteristic impedance. According to Equation 2.26, this impedance will be calculated as $Z_0 = 250.2114 \ \Omega$.

c.1 Lossless line

The lossless line is simulated in this part. Figure 2.30 is the PSCAD single-line diagram of the case study. The system circuit breaker closes at $t = 0.05$ sec. E_a and E_b are the sending end and receiving end voltages, respectively.

Figure 2.31 illustrates the sending end and receiving end voltages of the line. As it can be seen, it takes close to 2.5 ms for the wave to reach end of the line. In this case as it is discussed in this chapter, the reflection coefficient is zero. Therefore, voltage at the receiving end obtains the same value as the sending end voltage.

c.2 Lossy line

Figure 2.32 shows the sending end and receiving end voltages for the lossy case. Since the line is loaded with the surge impedance, there is no presentation of overvoltages at the end of the line. It is important to pay attention to the stairs illustrated in Figure 2.32. These stairs appear at the receiving end voltage due to the fact that the surge impedance (250.2114 Ω) calculated above is a little bit deviated from the actual surge impedance of the line. Therefore, the reflection coefficient is not exactly equal to zero, and there will be reflections at instants, which are independent of the traveling time (T). Resistance of the line attenuates the traveling wave, and after $3T$ s of switching, the receiving end voltage reaches the magnitude of the sending end voltage minus line voltage drop.

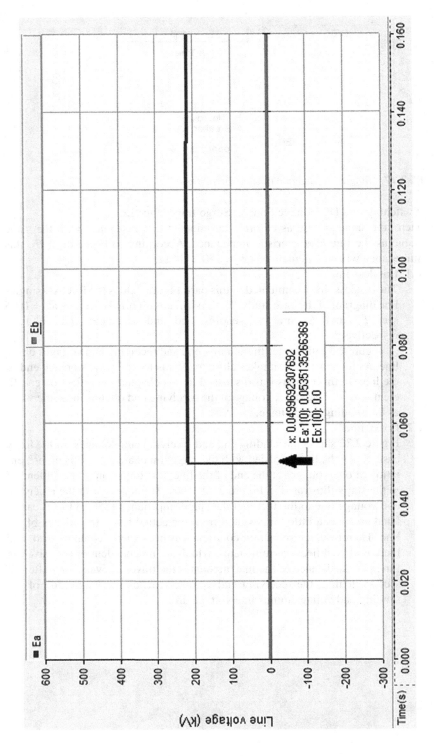

FIGURE 2.28 Voltages of the b.1 test.

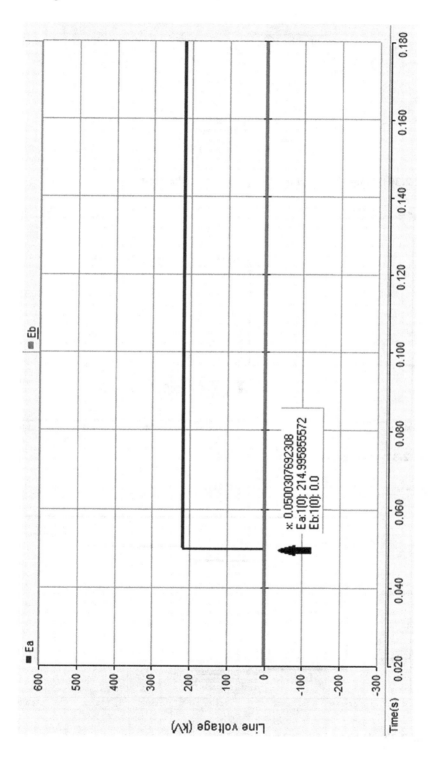

FIGURE 2.29 Voltages of the b.2 test.

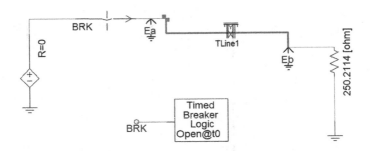

FIGURE 2.30 Single-line diagram of surge impedance load test.

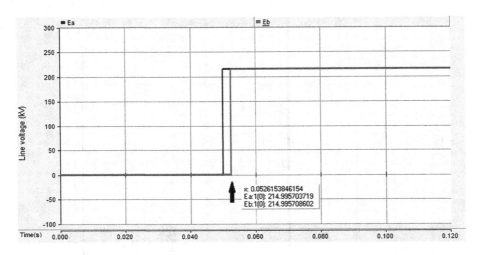

FIGURE 2.31 Voltages of the c.1 test.

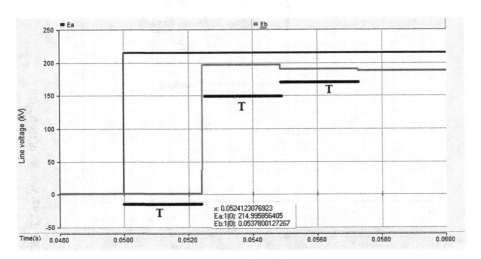

FIGURE 2.32 Voltages of the c.2 test.

Problems

2.1 What are the conditions that we need to consider in order to assume a constant speed for a traveling wave?

2.2 What are the conditions to be considered in order to have a traveling wave speed exactly the same as speed of light?

2.3 Following shows a voltage wave as a function of time and position, that is, $u(x,t)$.

$$u(x,t) = \text{Re}\left\{ \hat{u}_1 e^{-\gamma x} e^{j\omega t} + \hat{u}_2 e^{-\gamma x} e^{j\omega t} \right\} \text{ and } \gamma = \alpha + j\beta$$

Assuming this is a lossless line and voltage at the beginning of the line is $\hat{u}_0 \cos \omega t$ and voltage change per unit length is $-\beta \hat{u}_0 \sin \omega t$, find the function $u(x,t)$ in its simplest form, applying the initial condition to this equation.

2.4 Specify a condition where a system characteristic impedance is a constant number.

2.5 Consider Tables p2.1 and p2.2. In these tables, Z_1 and Z_2 are real and imaginary parts of the characteristic impedance. Both tables are specifying $\alpha, \beta,$ and $v\left(\dfrac{m}{s}\right)$ for an overheard line and a cable at different frequencies, respectively.

   a. Find low-, medium-, and high-frequency domains for the cable and the overhead line.

   b. Considering these two tables, specify if the propagation wave effect is greater in the cable or the overhead line.

TABLE P2.1

α, β, v Parameters, Real and Imaginary Parts of Characteristic Impedances (Z_1 and Z_2, respectively) of an Overhead Line as Functions of Frequency

f (Hz)	$\alpha \times 10^{-3}$ (1/km)	$\beta \times 10^{-3}$ (1/km)	$v\left(\dfrac{m}{\mu s}\right)$	Z_1 (Ω)	Z_2 (Ω)
0.1	0.1	0.0038	166	998	−51.00
1	0.104	0.0360	173	875	−242.00
10	0.155	0.2400	258	377	−170.00
20	0.170	0.4400	281	330	−99.00
50	0.176	1.0720	295	311	−42.00
100	0.177	2.1300	295	308	021.00
500	0.178	10.6200	295	307	−4.20
1000	0.178	21.2500	295	307	−2.13
2000	0.178	42.5000	295	307	−1.07

TABLE P2.2

α, β, ν **Parameters, Real and Imaginary Parts of Characteristic Impedances (Z_1 and Z_2, Respectively) of a Cable as Functions of Frequency**

f (Hz)	$\alpha \times 10^{-3}$ (1/km)	$\beta \times 10^{-3}$ (1/km)	$\nu \left(\dfrac{m}{\mu s} \right)$	$Z_1 (\Omega)$	$Z_2 (\Omega)$
1	1.43	0.225	27.9	7.16	1.12
10	2.08	1.553	40.4	10.5	7.60
20	2.70	2.387	52.6	13.9	11.51
50	3.98	4.045	77.6	21.4	18.58
100	5.34	6.037	104.0	30.6	25.47
500	8.94	18.030	174.0	71.7	35.00
1000	9.80	32.930	190.0	89.7	26.43
2000	10.12	63.670	197.0	98.4	15.50
10000	10.20	314.500	200.0	102.0	3.29

BIBLIOGRAPHY

1. Lou van der Sluis, *Transients in Power Systems*, John Wiley & Sons, 2001.
2. Hossein Mohseni, *Advanced High Voltage Engineering*, Tehran University Publications, 4th edition, 2018 (In Farsi).
3. Hermann W. Dommel, EMTP Theory Book, Microtran Power System Analysis Corporation, 1996.
4. Juan A. Martinez-Velasco, *Transient Analysis of Power Systems: A Practical Approach*, John Wiley & Sons, 2019.
5. Yoshihide Hase, *Handbook of Power System Engineering*, John Wiley & Sons, 2007.
6. Akihiro Ametani, Naoto Nagaoka, Yoshihiro Baba, and Teruo Ohno, *Power System Transients: Theory and Applications*, The United States of America: Taylor & Francis Group, 2013.
7. Zhengyou He, *Wavelet Analysis and Transient Signal Processing Applications for Power Systems*, John Wiley & Sons, 2016.
8. M. Mokhtari and G. B. Gharehpetian, *Grounding Systems Transients, Methodology and Modeling*, Book, LAP Lambert Academic Publishing, 2017, ISBN 978-3-659-88273-9.

3 Lattice Diagram and Its Applications

3.1 TRAVELING WAVE AS A FUNCTION OF TIME AND LOCATION

In order to calculate the magnitude of transient voltage at any point on the grid, a graphical approach has been developed that is called the lattice diagram. In this method, the wave is presented in the (x, t) plane. This is an old method without involving intensive calculations that renders accurate results. We will examine the methodology explained here with PSCAD modeling at the end of this chapter.

Assume we have the voltage wave presented in the plane of (u, t) or plane of (u, x), now, we eliminate the parameter u and will find a new presentation in the (x, t) plane. In Chapter 2, we saw that we can illustrate a wave as following:

$$u(x,t) = u_f(x - vt) + u_r(x + vt) \tag{3.1}$$

The first portion of this equation is a forward moving wave with the constant speed $(x = vt + a)$ and the next part is a backward moving wave with a constant speed $(x = -vt + b)$. Therefore, in the (x, t) plane, we will have $x \mp vt$; this presentation is a group of parallel lines that are shown in Figure 3.1. It is possible to cover all the possible points of the (x, t) plane with these lines, that is, we will obtain the magnitude of the voltage wave at any point of this plane.

In summary, the lattice diagram is utilized as a tool in finding the magnitude of a lossless traveling wave at different positions and different times. In this method, each transmission line is identified by characteristic impedance and propagation speed of the wave or the period of the propagation.

Period of propagation is defined as the time the wave needs in order to travel the entire length of the line.

$$T = l\sqrt{L'C'} \tag{3.2}$$

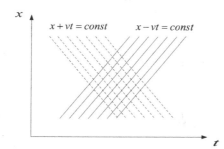

FIGURE 3.1 Traveling wave on a (x, t) plane.

DOI: 10.1201/9781003255130-3

Let us start with a simple example. Consider a transmission line with the characteristic impedance of Z_c. This line is connected to an electric source from one side and it is connected to the lumped impedance Z from the other side. The reflection coefficient of the wave at the end of the line can be found from Equation (3.3).

$$r = \frac{Z - Z_c}{Z + Z_c} \tag{3.3}$$

Now consider the following:

1. If the load impedance is really large ($z \to \infty$), the reflection factor (r) becomes 1.
2. If the load impedance is really small ($z \to 0$), the reflection factor (r) becomes 0.

Condition 1 as mentioned above, is the same as the open circuit line. In this condition, when the wave E reaches the end of the line, it will get reflected with the same magnitude. When the reflected wave reaches the beginning of the line, because the source impedance is zero and will get reflected again with the same magnitude but the opposite sign. This is illustrated in Figure 3.2. It will take T seconds for the wave

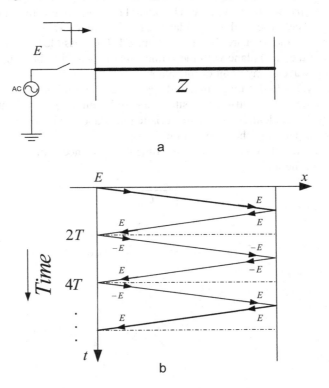

FIGURE 3.2   (a) Open circuit line. (b) Its lattice diagram.

to travel from the source to the end of the line. Therefore, for all $t < T$, the voltage at the end of the line is zero. Using the reflection factor after the time T, voltage at the end of the line is $2E$. It takes $2T$ seconds for the wave to reach the source again. Now, when the wave reaches the end of the line again, the voltage at the end point becomes zero as the voltage at $t = T$ was $2E$ and now the voltage $-2E$ at $t = 3T$ has reached the end of the line. Therefore, the sum of $2E+(-2E)$ will give the voltage zero at the end of the line. The situation for sending of the line is governed by the zero impedance of the source. Since the source impedance is zero, always the same wave with a negative sign gets reflected. Therefore, the voltage at the sending end of the line at any time is E.

The voltage diagram of the sending end and receiving end of the line is illustrated in Figure 3.3. The voltage at the sending end does not change, but at the receiving end, the voltage oscillates between 0 and $2E$.

Now we start looking at the short circuit condition. This case is shown in Figure 3.4a. In this condition, when the wave reaches the end of the line (short circuit), it will get reflected with the same magnitude but a negative sign. Therefore, at $t = T$, the coming wave and the reflected wave will cancel each other. When the reflected wave $-E$ reaches the sending end of the line, due to the zero impedance of the source, it will get reflected back with the same magnitude and an opposite sign. Therefore, magnitude of the voltage at sending end of the line is E at any time. This condition is illustrated in Figure 3.4b. Using the lattice diagram, we can find voltage and current at each point of transmission line. Assume we need to find the voltage at the point x from the sending end. If the time required for the wave to reach to the end of the line is T, therefore, it takes $(x/L)T$, for the wave to reach the point x, given L is

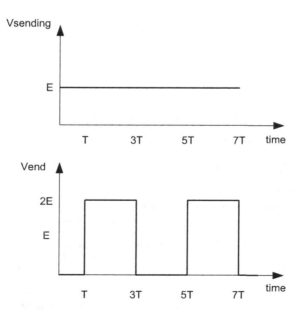

FIGURE 3.3 Voltage at sending end and receiving end of open circuit line.

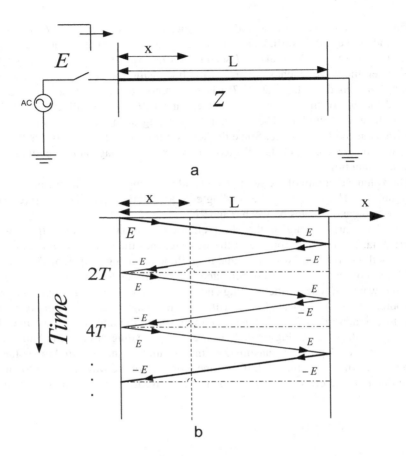

FIGURE 3.4   (a) Short circuit diagram. (b) Lattice diagram.

the length of the line. When the wave reached the point x, voltage at that point becomes the same as the source voltage. On the way back, at $t = 2\left(\dfrac{1-x}{L}\right)T$, the $-E$ wave reaches the point x again. This will cause the voltage at that point to become zero at the specified time stamp.

Figure 3.5 Voltage magnitude at point x as a function time. Current wave similar to the voltage wave reaches the point x at $t = (x/L)T$. Amplitude of the current wave is $I = \dfrac{E}{\pm Z}$. The impedance Z takes a positive or a negative sign according to the sign of the traveling wave. The reflected wave is $I' = -\dfrac{E}{-Z} = \dfrac{E}{Z} = I$. Therefore, when the wave reaches the point x for the second time (i.e., on the way back), the current wave at that point is the sum of all the currents at that point and it will be $2I$. This amplification continues and has been illustrated in Figure 3.5.

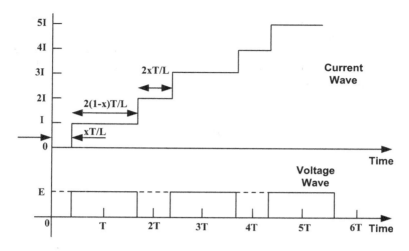

FIGURE 3.5 Current and voltage waves at point x.

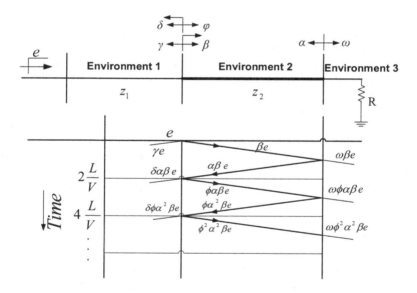

FIGURE 3.6 Lattice diagram for a complex system.

What has been discussed so far are simple examples of utilization of lattice diagram. In order to analyze a more general condition including losses in the line, we consider cases such as cable connection and resistance in the load. In time, the number of steps in the lattice diagram will increase. Figure 3.6, illustrates a system comprised of three propagation environments. The first environment is a transmission and the second is a cable. At $t = 0$, the wave reaches the point where there is an alteration in the propagation environment (border condition). This figure contains break and reflection coefficient. Here, α is the reflection coefficient in the second environment, δ is the reflection coefficient in the first environment, ϕ is the passing coefficient into the second environment, and ω is the passing coefficient into the third environment. The negative slope in the lattice diagram is a constant number. These propagation parameters are formulated as following:

$$\alpha = \frac{R - Z_2}{R + Z_2} \tag{3.5}$$

$$\omega = \frac{2R}{R + Z_2} \tag{3.6}$$

$$\delta = \frac{2Z_1}{Z_1 + Z_2} \tag{3.7}$$

$$\phi = \frac{Z_1 - Z_2}{Z_1 + Z_2} \tag{3.8}$$

In Figure 3.6, the wave (e) reaches the end of the line in the first environment and (βe) enters the environment 2, (γe) gets reflected back to environment 2. It is possible to calculate γ from propagation equations. The wave that reaches the end of the environment 3 gets reflected back to environment 2 by the coefficient (βe). When this wave hits the third environment $(\omega \beta e)$, it enters the third environment and $(\alpha \beta e)$ gets reflected back to the second environment. When the wave $(\alpha \beta e)$ reaches the first environment $(\delta \alpha \beta e)$, it enters the first environment and $(\phi \alpha \beta e)$ will be reflected back to the second environment. The process continues and voltage at each point can be calculated by adding the voltage values associated with one position at different times. For instance, the voltage at the point where the transmission line is connected to the cable can be calculated using Equation (3.8):

$$e_T = e[\beta + \delta\alpha\beta U(t - 2T) + \delta\phi\alpha^2\beta U(t - 4T) + \delta\phi^2\alpha^3\beta U(t - 6T) + \cdots] \tag{3.8}$$

In this equation, $U(t)$ is the unit step function, T is the time that it takes for the wave to travel through the second environment, and it is assumed that the length of the first environment is considered to be infinite, in order for eliminating the possibility of the reflection from this environment. T can be found as:

$$T = \frac{L}{v} \tag{3.9}$$

In order to calculate the voltage at the beginning of the cable, it must be said that $t = 2T$ is the time required for the wave to travel through the cable and gets fully reflected with respect to time.

3.2 CABLE CONNECTION TO TRANSMISSION LINE

Assume that there is a need for energy transmission to a point in a city. Common practice standards limit the installation of transmission lines in cities. Therefore, a portion of the electric path is going to be a cable. At the point where the transmission line is connected to the cable, there will be a change in the propagation environment that the wave travels to. This condition is illustrated in Figure 3.7.

In this figure, the transmission line is connected to the cable at point a. Figure 3.8 shows that the wave u_{f1} reaches the borderline between propagation environment 1 and 2, u_{f2} enters the environment 2, and u_{r1} gets reflected back to environment 1.

Considering Figures 3.7 and 3.8, we can write the following equations for the connection point a.

$$u_{f1} + u_{r1} = u_{f2} \tag{3.10}$$

$$i_{f1} + i_{r1} = i_{f2} \tag{3.11}$$

Using the relationship between traveling voltages and currents and Equation (3.11), we have:

$$\frac{u_{f1}}{Z_1} - \frac{u_{r1}}{Z_1} = \frac{u_{f2}}{Z_2} \tag{3.12}$$

Dividing Equation (3.10) by Z_1 and adding it to Equation (3.12), we will have:

$$2u_{f1} = u_{f2}\left(1 + \frac{Z_1}{Z_2}\right) \tag{3.13}$$

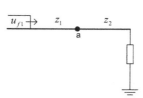

FIGURE 3.7 Wave propagation in a system of transmission line and cable.

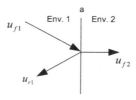

FIGURE 3.8 Passing and reflection of a traveling wave in two environments.

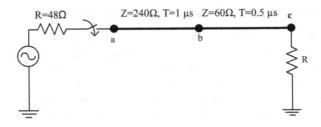

FIGURE 3.9   Source connection to the line and cable.

$$u_{f_2} = \frac{2Z_2}{Z_1 + Z_2} u_{f_1} \tag{3.14}$$

Example: Figure 3.9 shows an alternative source with the magnitude of 150 kV and the internal resistance of 48 Ω. This source gets connected to an overhead line with the impedance of 240 Ω and $T = 1\mu s$. At the end of the line, there is a cable with the impedance of 60 Ω and $T = 0.5\,\mu s$, which supplies the 60 Ω load. Find the sending end and receiving end voltages.

Before showing the solution, we need to consider the fact that for an overhead line L' has a high value and C' has a low value. Therefore, considering the real dimensions of an overhead line, the value for Z_c is 200–300 Ω $\left(Z_c = \sqrt{\dfrac{L'}{C'}} \right)$. For underground cables, the amount of capacitance per length is high and the amount of inductance per length is low. Therefore, the typical value for Z_c is smaller and it is around 30–60 Ω. Now for this example when the breaker is closed, at point a we will have:

$$u_o = \frac{240}{240 + 48} \times 150 = 125\,\text{kV}$$

For the wave that gets propagated from a to b, we have:

$$r_b = \frac{60 - 240}{60 + 240} = -0/6 \Rightarrow b_b = 0/4$$

At point c, we will have:

$$r_c = 0$$

For the wave that travels from b to a, we have:

$$r_a = \frac{48 - 240}{48 + 240} = -\frac{2}{3}$$

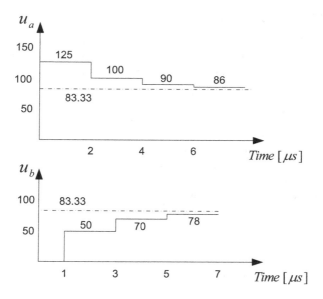

FIGURE 3.10 Voltage in kV at points a and b.

Now we start with the lattice diagram to show the magnitude of voltage at points a and b. This lattice diagram has been presented in Figure 3.10.

In the lattice diagram, the system is considered to be lossless. Therefore, with time, the diagrams associated with points a, b, and c will merge into a certain number. We can see that voltage at point b, which is the entrance of the cable, changes in time. In other words, at this point, we will have a voltage stress $\left(\text{in } \dfrac{kV}{\mu s}\right)$. This factor is an important factor in designing the insulation. Considering the insulation matters, beginning of the cable is the weakest point in transients.

In solving the problem, we considered a constant magnitude for the source, but as it is mentioned, the source is an alternating source. The speed of wave propagation is really high so making the assumption that the source has a fixed value does not cause large errors in calculations. If the length of the line is 1500 m and the wave speed being the same as speed of light, therefore, the time for a back and forth travel in the line will calculated as 10 μs. The following calculation shows that the change in the sine or cosine waveform in 10 μs (e.g., around $t = 0$) is a very small number:

$$V\cos(2\pi \times 50 \times t) - V\cos(2\pi \times 50 \times (t + 10 \times 10^{-6}))\big|_{t=0} = 1.5 \times 10^{-9} \times V \cong 0$$

Therefore, in an analysis related to traveling waves and their impacts, we can consider a fixed magnitude for alternating sources (synchronous generators).

3.3 CLOSING RESISTANCE CIRCUIT BREAKERS

In this part of the chapter, we study the transients related to the closing of circuit breakers. At the energization time, the system state changes, and after a certain time, the transients will go away and the system enters a steady-state condition.

3.3.1 SOURCE INTERNAL RESISTANCE

In this part, we consider a source that has an internal resistance. Assume an open circuit line is going to be stressed with the mentioned source. In studies similar to the this (i.e., line energization), the network beyond the line can be modeled as a source and internal resistance R_i (or impedance). Figure 3.11 shows this system.

The forward traveling current wave will be calculated as:

$$i_f = \frac{u_f}{z} \tag{3.16}$$

On the contrary, in the lumped circuit including u and R_i, we will have:

$$u_f = u - R_i i_f \Rightarrow u = u_f + R_i i_f = R_i \left(\frac{u_f}{z} \right) + u_f \tag{3.17}$$

In this equation, u_f is the voltage at the source terminal (i.e., after internal resistance). U is the source voltage and Z is the characteristic impedance of the line. Let us assume the internal resistance of the source is the same as the characteristic impedance of line (i.e., $R_i \approx Z$); therefore, we will have:

$$u_f = \frac{u}{2} \tag{3.18}$$

According to the info gathered in Chapter 2, voltage at the end of an open circuit line is twice the entrance voltage.

$$u_{\text{end}} = 2u_f = u \tag{3.19}$$

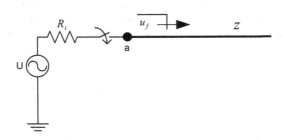

FIGURE 3.11 Cable energization.

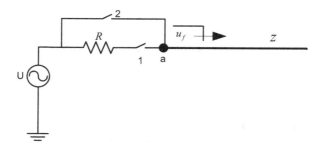

FIGURE 3.12 Closing breaker in line energization.

Therefore, considering the equations above with smaller source resistance, we will have higher u_f and therefore the voltage at the end of the line gets higher. In order to decrease the voltage at the end of the line and control the possible insulation stresses, a series closing resistance will get connected to the source before energization of the line. It is possible to control the voltage at the end of the line. Keeping the resistance in the circuit in steady-state conditions will cause unwanted voltage drops and high losses.

Figure 3.12 shows a cable that is being energized with a closing resistance. As this figure suggests, the process of line energization starts with closing the breaker 1. The total resistance after the voltage source can be calculated by adding the internal resistance of the source to the closing resistance. After some time, when the transients are damped, breaker 2 closes and R will be eliminated from the circuit to prevent unwanted losses. This way, the voltage that reaches end of the line is controlled and the probability of damaging the insulation at the end of line is decreased.

3.3.2 DEAD LINE CONNECTION TO LIVE LINE

Figure 3.13 shows a condition that a live line is getting connected to a dead line. If E_{10} is the initial line voltage before switching and E_{20} is the initial line voltage of the dead line and E_{12} is the voltage difference on the breaker, we will have:

$$E_{1o} = E, E_{2o} = 0 \tag{3.20}$$

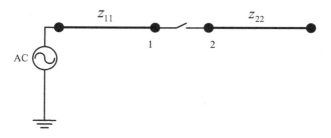

FIGURE 3.13 Dead line charge through a live line.

$$E_{12} = E_{1o} - E_{2o} = E \tag{3.21}$$

Eventually, when the system reaches a steady state, the voltage across the breaker becomes zero:

$$E_{12} = 0 \tag{3.22}$$

With closing the breaker, the current I starts flowing through the circuit. This will resemble the situation where voltage E is applied directly on the breaker. The wave E will see the two impedances of Z_{11} and Z_{22}, which are the line impedance for live and dead line, respectively. We can calculate the current I as follows:

$$I = \frac{E}{Z_{11} + Z_{22}} \tag{3.23}$$

When the current I flows through the line, two voltage waves with amplitudes, as shown below, start traveling through both lines.

$$E_1 = -Z_{11}I = \frac{-Z_{11}}{Z_{11} + Z_{22}} E \tag{3.24}$$

$$E_2 = Z_{22}I = \frac{Z_{22}}{Z_{11} + Z_{22}} E \tag{3.25}$$

In the above equations, E_1 is the voltage traveling back to the live line and E_2 is the forward moving wave that travels through the dead line. Considering the lattice diagram in Figure 3.14, the backward wave will be $E + E_1$ in total and the forward moving wave is $(0 + E_2)$. In case of having equal line impedances, voltage at the

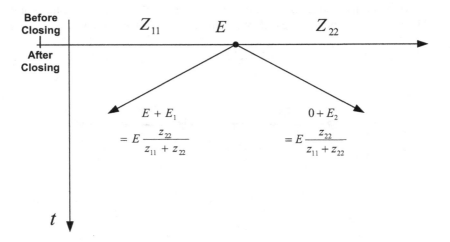

FIGURE 3.14 Forward and backward waves after breaker closing.

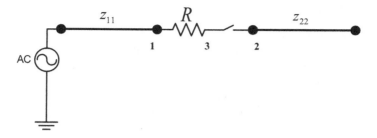

FIGURE 3.15 Closing resistance to energize a line through a live line.

breaker (E) will be divided into two equal waves $E/2$ and one travels to the right and the other one travels to the left.

If a closing resistance is used, the circuit will be changed to the one shown in Figure 3.15.

Voltages at points 1, 2, and 3 can be found as following:

$$E_{10} = E, E_{30} = E, E_{20} = 0 \qquad (3.26)$$

The index 0 refers to the initial time before breaker closing. As it can be seen in the figure, point 3 is obtained after the closing resistance and before the breaker, and we have:

$$I = \frac{E}{(R + Z_{11}) + Z_{22}} \qquad (3.27)$$

$$E_2 = Z_{22}I = \frac{Z_{22}}{R + Z_{11} + Z_{22}} E \qquad (3.28)$$

E_2 is the propagated wave through the line with characteristic impedance of Z_{22} and E_1 is the propagated wave to the line with the characteristic impedance of Z_{11}, and E_R is the voltage across the resistance R.

$$E_1 = -Z_{11}I = \frac{-Z_{11}}{R + Z_{11} + Z_{22}} E \qquad (3.29)$$

Therefore, we have:

$$E_R = RI = \frac{R}{R + Z_{11} + Z_{22}} E \qquad (3.30)$$

As it can be seen in the above equations, the closing resistance will cause a decrease in E_2 (traveling wave to the dead line). If the line is open circuited, utilization of closing resistance will decrease the magnitude of the transient overvoltage at the end of the line. The resistance is modeled as lumped sum circuit element; therefore, the time that takes for the wave to travel through this element is zero.

3.4 REFLECTION IN BRANCHES

Assume the wave U_f is traveling through a line with the characteristic impedance of Z. Figure 3.16 shows that this line is being connected to multiple lines with characteristic impedances as $(Z_1, Z_2, \ldots, Z_n)$. This case is similar to a configuration, which can be seen in transmission substations.

When the wave reaches the branching point, electric charges will be distributed into the branches; therefore, at this point, we will have:

$$i_f + i_r = i_{f_1} + i_{f_2} + \cdots + i_{f_n} \tag{3.31}$$

The voltage wave at the branching spot will get applied to all the branches:

$$u_f + u_r = u_{f_1} = u_{f_2} = \cdots = u_{f_n} \tag{3.32}$$

Using Equation (3.31) and the relationship between voltage and current waves, we will have:

$$\frac{u_f}{z} - \frac{u_r}{z} = \frac{u_{f_1}}{z_1} + \frac{u_{f_2}}{z_2} + \cdots + \frac{u_{f_n}}{z_n} \tag{3.33}$$

$$2u_f = u_{f_1}\left(1 + \frac{z}{z_1} + \frac{z}{z_2} + \cdots + \frac{z}{z_n}\right) \tag{3.34}$$

$$2u_f = u_{f_1}\left(1 + z\sum_i \frac{1}{z_i}\right) \tag{3.35}$$

According the last equation above, the line impedances run in parallel.

$$\sum \frac{1}{Z_i} = \frac{1}{Z_g} \tag{3.36}$$

The impedance Z_g in the equation above is the equivalent impedance of all the branches. Therefore, all the branches are equalized as one branch. Therefore, the equations developed previously will become valid for this case as well. Therefore, for the forward traveling and backward traveling waves, we will have:

$$u_{f_1} = \frac{2Z_g}{Z + Z_g}u_f \tag{3.37}$$

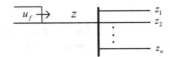

FIGURE 3.16 Reflection in branches.

TABLE 3.1
Break and Reflect Coefficients

N	B	r
0	2	1
1	82/1	82/0
10	1	0
100	18/0	82/0−
∞	0	1−

$$u_r = u_{f_1} - u_f = \frac{Z_g - Z}{Z + Z_g} u_f \qquad (3.38)$$

Table 3.1 shows values for break (b) and reflect (r) coefficients for multiple lines in parallel. The table is made by making the assumption that a wave is traveling to a line with a characteristic impedance of 50 Ω and it reaches n branches that each has 500 Ω, characteristic impedance.

3.5 SPECIAL CASES IN REFLECTION

3.5.1 Capacitor Connection at the End of the Line

Voltage drop exists for the lines that deliver load to far away distances. If the voltage drop falls below 95% of the nominal voltage, capacitor banks are installed to boost the voltage. The capacitor bank is responsible for providing reactive power support. Let us study a line that is terminated to a capacitor. Voltage wave u_f is propagated to the line as it is shown in Figure 3.17.

We assume the starting point of the study is when the wave reaches the capacitor. As it is explained earlier in the following equations, r and f represent reflected and fractured waves, respectively.

$$i_c = i_f + i_r \qquad (3.39)$$

$$u_c = u_f + u_r \qquad (3.40)$$

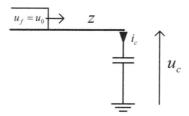

FIGURE 3.17 Traveling wave in a line with capacitor termination.

$$i_f = \frac{u_f}{Z} \tag{3.41}$$

$$i_r = -\frac{u_r}{Z} \tag{3.42}$$

$$i_c = C\frac{du_c}{dt} \tag{3.43}$$

$$i_c = i_f + i_r = C\frac{du_c}{dt} = \frac{u_f}{Z} - \frac{u_r}{Z} \tag{3.44}$$

$$C\left(\frac{du_f}{dt} + \frac{du_r}{dt}\right) = \frac{u_f}{Z} - \frac{u_r}{Z} \tag{3.45}$$

$$u_f - u_r = CZ\frac{du_f}{dt} + CZ\frac{du_r}{dt} \tag{3.46}$$

For a small number of ε that is greater than zero, we know $u_f = u_0$ and $\frac{du_f}{dt} = \frac{du_0}{dt} = 0$, therefore:

$$u_o - u_r = CZ\frac{du_r}{dt} \tag{3.47}$$

The solution for the above first-order differential equation is an exponential function. We find the parameter m by solving the characteristic equation of the differential Equation (3.47)

$$m = -\frac{1}{ZC} \tag{3.48}$$

$$u_r(t) = Ae^{-t/cz} + B \tag{3.49}$$

In order to find the coefficients A and B, we will need to apply the initial conditions. At $t = 0$, the capacitor will act as a short circuit; therefore, we have:

$$u_r(0) = A + B = -u_o \tag{3.50}$$

The negative sign come from the fact that Equation (3.40) is zero at $t = 0$ and $u_f = u_0$. After a long time when the capacitor is fully charged, it acts as an open circuit.

$$u_r(\infty) = B = u_o \Rightarrow \begin{cases} A = -2u_o \\ B = u_o \end{cases} \tag{3.51}$$

Therefore, the reflected voltage can be written as:

$$u_r(t) = u_o \left(1 - 2e^{-t/cz} \right) \qquad (3.52)$$

Considering Equation (3.40) again, we will have:

$$u_c = u_f + u_r = u_r(t) = 2u_o \left(1 - 2e^{-t/cz} \right) \qquad (3.53)$$

The associated waves are plotted in Figure 3.18 a and b. In the above equations, C can represent a capacitor bank or a model of a no-load cable or a model for a no-load

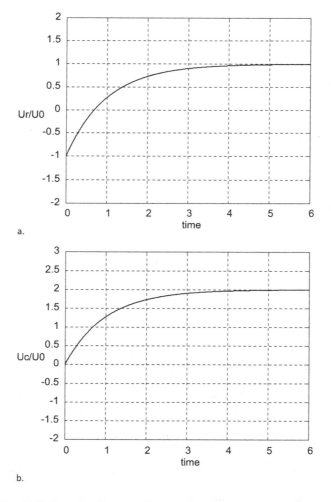

a.

b.

FIGURE 3.18 (a) Reflected voltage. (b) Capacitor voltage.

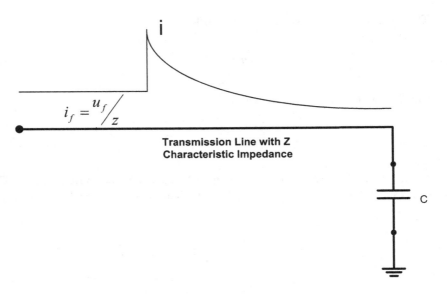

FIGURE 3.19   Current equivalent wave in a transmission line.

transformer for transient studies. Considering Equation (3.53), we can find the capacitor's current as given by Equation (3.54) and the reflected current as given by Equation (3.57) and shown in Figure 3.19.

$$i_c = c\frac{du_c}{dt} = \frac{2u_0}{z}e^{-t/cz} = 2i_f e^{-t/cz} = 2\frac{u_f}{z}e^{-t/cz} \tag{3.54}$$

$$i_c = i_f + i_r \tag{3.55}$$

$$i_r = i_c - i_f \tag{3.56}$$

$$i_r = i_f\left(2e^{\frac{-t}{cz}} - 1\right) \tag{3.57}$$

From the last equation above, it is possible to conclude that high voltage capacitors that are exposed to a transient wave will experience an impulse current. Another application for high voltage capacitors is installing them in parallel to a line arrester; this can be concluded from Figure 3.18b. It can be seen in this figure that the capacitor can prevent any sudden change (in this case, a step change) and reduce the voltage stress $\left(\dfrac{kV}{\mu s}\right)$ at the connection point. Figure 3.20 illustrates characteristic of line arresters. The stress $\dfrac{kV}{\mu s}$ (rate of change of voltage in time) is very high in transient waves. The arrester characteristic falls below the peak of the wave. Therefore, the

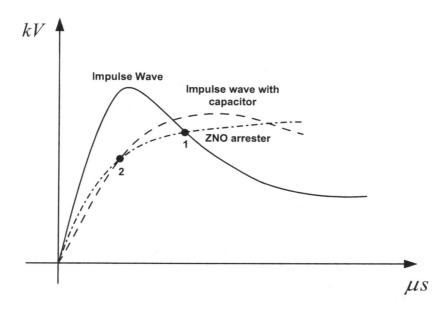

FIGURE 3.20 Arrester function with and without capacitor.

arrester does not operate on-time for the peak of the voltage and that causes damage to the equipment protected by the arrester.

As it can be seen in Figure 3.20, the first point where the arrester operates (point 1) is after the peak of the transient wave. System equipment will tolerate the peak of the wave without arrester operation and this can be harmful to the equipment. In order to prevent the abovementioned issue, a combination of capacitor bank and an arrester will be used. We can prove that the rate of change of voltage $\left(\dfrac{du}{dt} \right)$ can be decreased when the capacitor is installed. In a case wherein there is a capacitor installed in parallel with the arrester, the impulse wave has a peak less than the case without the capacitor and point 2, which is below the peak of the wave, is where the arrester operates. Therefore, the abovementioned combination will secure system equipment that are exposed to transients. In fact, the system can experience an impulse of $2u_0$ without capacitor installation, which is the case for an open circuit line.

3.5.2 Inductors Connection at the End of the Line

In transmission-level substations and, specially, the ones in which long lines are connected to them, in low load conditions, voltage shows an unfavorable rise (the Ferranti effect). In order to alleviate this dangerous rise in voltage, a reactor, which can be switched at the end of line, is used. In the event of voltage rise, the reactors will switch into the circuit, and with reactive power consumption, they prevent the unfavorable rise of the voltage. Also, in long transmission lines, capacitive effect

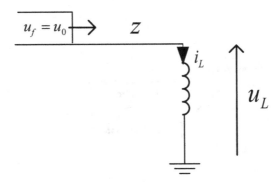

FIGURE 3.21 Reactor at the end of the line.

can cause voltage rise at the end of the transmission line. In order to control the rise in voltage, reactors with a disconnector (no load breakers) are utilized to get connected to the end of the transmission line. In this case, the result of economic studies has shown that using a circuit breaker is not economic. These reactors are usually switched once in each season, depending on the grid load. These reactors are connected/disconnected to/from the line after line de-energization (under no load condition). This section will study transients of these reactors (Figure 3.21).

We can write the following equations for transient voltage and current waves at the end of the line.

$$\begin{cases} i_L = i_f + i_r = \dfrac{u_f}{Z} - \dfrac{u_r}{Z} \\ u_L = u_f + u_r = L\dfrac{di_L}{dt} \end{cases} \tag{3.58}$$

The voltage across the inductor will be calculated from the forward and backward traveling waves at the end of the line. First, we change the voltage equations to current equations.

$$u_L = u_f + u_r = Zi_f - Zi_r = L\frac{di_L}{dt} \tag{3.59}$$

Now, we use $i_f + i_r$ instead of i_L and we will have:

$$L\frac{di_f}{dt} + L\frac{di_r}{dt} = Zi_f - Zi_r \Rightarrow i_f - i_r = \frac{L}{Z}\left(\frac{di_f}{dt} + \frac{di_r}{dt}\right) \tag{3.60}$$

Considering the fact that u_f is a step wave and $i_f = \dfrac{u_f}{Z}$, we can conclude that i_f is a step wave as well. Therefore, the derivative of that with respect to time is 0 at the time, which is epsilon greater than 0. Now, we change the equations as below:

$$\frac{di_f}{dt} = 0 \Rightarrow i_f = \frac{V_o}{Z} = i_r + \frac{L}{Z}\frac{di_r}{dt} \tag{3.61}$$

The abovementioned differential equation is a first-order equation. The solution for this equation is an exponential function. Forming the characteristic equation will give us the following solution:

$$m = -\frac{Z}{L} \tag{3.62}$$

$$i_r = Ae^{\frac{-tZ}{L}} + B \tag{3.63}$$

With the knowledge that the inductor will act as an open circuit at the start and short circuit after a long time, we can find the coefficients A and B presented in the latest equation.

$$\begin{cases} t = 0 \rightarrow i_L = 0 \rightarrow i_r(0) = -i_f \\ t = \infty \rightarrow i_r = i_f = \dfrac{V_o}{Z} \end{cases} \tag{3.64}$$

Placing the solutions for A and B, will alter Equation (3.63) as:

$$i_r = \frac{V_o}{Z}\left(1 - 2e^{\frac{-tZ}{L}}\right) \tag{3.65}$$

Using Equation (3.65), the voltage and current equations can be rewritten as follows:

$$\begin{cases} i_L = \dfrac{2V_o}{Z}\left(1 - 2e^{\frac{-tZ}{L}}\right) \\ V_L = \dfrac{2V_o}{Z}e^{\frac{-tZ}{L}} \end{cases} \tag{3.66}$$

Figures 3.22 and 3.23 are illustrating the diagrams for Equation (3.66). As it can be seen that when the transient voltage wave reaches the inductor, the equipment insulator faces a large electric tension. This is true for equipment such as transformers, reactors, and induction motors. The design of insulation of these devices must consider this possible overvoltage.

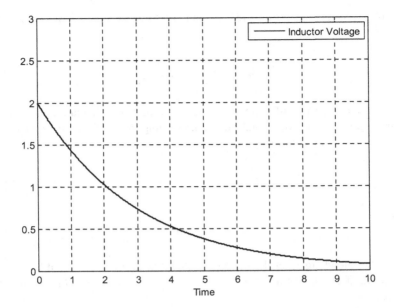

FIGURE 3.22   Inductor voltage.

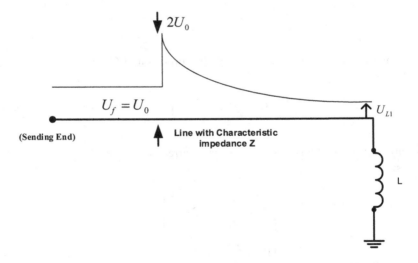

FIGURE 3.23   Equivalent voltage and algebraic sum of the forward and backward moving waves (dual for the current of the capacitor installed at the end of the line).

3.5.3 Parallel LC Connection at the End of the Line

In case of transients, it is not possible to model the coils as only inductors. When the traveling waves reach these coils, small capacitors will come into play. These capacitors can be considered parallel with the inductors. In reality, high-frequency transients with a limited speed will cause a voltage distribution with respect to the

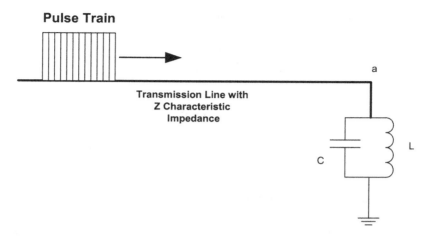

FIGURE 3.24   Parallel capacitor and inductor at the end of the line.

ground in internal parts of the coil. This effect in modeling will be represented as capacitors. In transformer winding modeling, the capacity of the mentioned capacitor is in nF range and the inductor is in mH range (Figure 3.24).

We can write the circuit equations as below:

$$u_a = u_f + u_r \tag{3.67}$$

$$i = i_C + i_L = i_f + i_r \tag{3.68}$$

$$i = C\frac{du}{dt} + \frac{1}{L}\int u\,dt \tag{3.69}$$

Taking a derivative of the last equation will give:

$$\frac{di}{dt} = C\frac{d^2u}{dt^2} + \frac{u}{L} \tag{3.70}$$

Using the current equations, we will have:

$$\begin{cases} \dfrac{di_f}{dt} + \dfrac{di_r}{dt} = C\dfrac{d^2u}{dt^2} + \dfrac{u}{L} \\[2mm] i_r = \dfrac{-u_r}{Z} \\[2mm] i_f = \dfrac{u_f}{Z} \\[2mm] \dfrac{di_f}{dt} = 0 \end{cases} \Rightarrow C\frac{d^2u}{dt^2} + \frac{1}{Z}\frac{du}{dt} + \frac{u}{L} = 0 \tag{3.71}$$

The solution for the second order differential equation will be:

$$\frac{d^2u}{dt^2} + \frac{1}{CZ}\frac{du}{dt} + \frac{u}{LC} = 0 \tag{3.72}$$

$$m^2 + \frac{m}{CZ} + \frac{1}{LC} = 0 \tag{3.73}$$

$$m = \frac{-1}{2CZ} \pm \sqrt{\frac{1}{4C^2Z^2} - \frac{1}{LC}} = -\alpha \pm \gamma \tag{3.74}$$

If γ is a real number, the differential equation will be solved as:

$$u(t) = Ae^{(-\alpha+\gamma)t} + Be^{(-\alpha-\gamma)t} = (Ae^{\gamma t} + Be^{-\gamma t})e^{-\alpha t} \tag{3.75}$$

Now, let us see how valid the assumption of γ a real number is. If γ is an imaginary number, then we should have:

$$\left(\frac{1}{2cz}\right)^2 < \frac{1}{Lc} \Rightarrow L < 4cz^2 \tag{3.76}$$

This formula is only valid if C is large and L is small. As we described above, the inductor is larger than the capacitor for a real case; this proves that γ cannot be an imaginary number and it is a real number.

At $t = 0$, that is, the instant of the voltage wave entrance into the line end, and considering the presence of the capacitor as a short circuit, the voltage is zero ($u(0) = 0$), Therefore, we have:

$$u(0) = 0 \Rightarrow B = -A \tag{3.77}$$

$$u(t) = A(e^{\gamma t} - e^{-\gamma t})e^{-\alpha t} \tag{3.78}$$

$$u(0) = 0 \Rightarrow u_r(0) = -u_f \tag{3.79}$$

$$i_r(0) = \frac{-u_r(0)}{Z} = \frac{u_f}{Z} = i_f \tag{3.80}$$

$$i(0) = i_f + i_r(0) = 2i_f = 2\frac{u_f}{Z} \tag{3.81}$$

$$i(0) = i_L(0) + i_C(0) \Rightarrow i_C(0) = i(0) = 2\frac{u_f}{Z} \tag{3.82}$$

Also, at $t = 0$, the inductor will present an open circuit connection. Therefore, $i_L(0) = 0$, and we have:

$$i_C(0) = C\frac{du(0)}{dt} \Rightarrow \frac{du(0)}{dt} = 2\frac{u_f}{CZ} \tag{3.83}$$

On the other hand, we have:

$$\frac{du(0)}{dt} = 2A\gamma = 2\frac{u_f}{CZ} \Rightarrow A = \frac{u_f}{CZ\gamma} \tag{3.84}$$

$$u(t) = 2u_f\left(\frac{\alpha}{\gamma}\right)(e^{\gamma t} - e^{-\gamma t})e^{-\alpha t} \tag{3.85}$$

Figure 3.25 shows a comparison of the voltage at the end of the line, when we have a parallel case of an inductor and a capacitor with only the inductor. As it can be seen at $t = 0$, when the wave reaches the parallel LC circuit, the voltage is similar to the condition where there is only a capacitor connected at the end of the line (i.e., zero voltage); as the time passes by, the wave becomes similar to the voltage wave where there is only an inductor connected to the line. This shows the fact that modeling and considering correct values for capacitors when dealing with transients is a crucial matter.

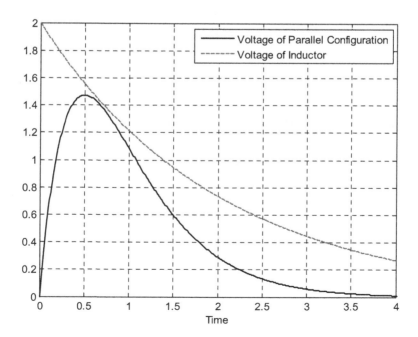

FIGURE 3.25 Voltage comparison of two cases.

3.5.4 Series LC Connection at the End of the Line

In modeling the equipment for high voltage transients, it is possible to have cases of series capacitors and inductors. There is rarely an intentional series connection of an inductor and a capacitor in power systems. In the case of this series connection, there is a danger of intense resonance. This resonance can cause irreversible damage to the exposed equipment. The series LC connection can represent and model transient behavior of some equipment.

In Figure 3.26, a is the end of the line and b is the connection point of the capacitor and inductor.

$$u_a = u_f + u_r = L\frac{di}{dt} + u_b \tag{3.86}$$

$$i = i_f + i_r = c\frac{du_b}{dt} \tag{3.87}$$

Taking derivatives will give:

$$\frac{di}{dt} = c\frac{d^2u_b}{dt^2} \tag{3.88}$$

Replacing Equation (3.88) with (3.86), we will have:

$$u_a = u_b + LC\frac{d^2u_b}{dt^2} \tag{3.89}$$

$$u_f + u_r = u_b + LC\frac{d^2u_b}{dt^2} \tag{3.90}$$

Multiplying Equation (3.87) by Z, we will have:

$$u_f - u_r = ZC\frac{du_b}{dt} \tag{3.91}$$

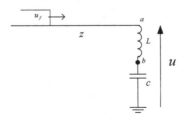

FIGURE 3.26 Series capacitor and inductor at the end of the line.

Adding Equation (3.90) to (3.91), we will have:

$$2u_f = LC\frac{d^2u_b}{dt^2} + ZC\frac{du_b}{dt} \qquad (3.92)$$

Same as before, the solution of the differential equation will be calculated by forming the characteristic equation:

$$LCm^2 + ZCm + 1 = 0 \qquad (3.93)$$

$$m = -\frac{Z}{2L} \pm \sqrt{\left(\frac{Z}{2L}\right)^2 - \frac{1}{LC}} \qquad (3.94)$$

Considering $\left(\dfrac{Z}{2L}\right)^2 < \dfrac{1}{LC}$, we will have:

$$\left(\frac{z}{2L}\right)^2 < \frac{1}{Lc} \Rightarrow m = -\alpha \pm j\gamma \qquad (3.95)$$

$$u_b(t) = e^{-\alpha t}(c_1\cos(\gamma t) + c_2\sin(\gamma t)) + 2u_f \qquad (3.96)$$

As usual, we use the borderline conditions of $t = 0$ and $t \to \infty$ to find c_1 and c_2.

$$u_b(0) = 0 \Rightarrow c_1 = -2u_f \qquad (3.97)$$

$$i_c(0) = 0 \Rightarrow C\frac{du_b}{dt} = 0 \Rightarrow c_2 = -2u_f\frac{\alpha}{\gamma} \qquad (3.98)$$

Therefore, we have:

$$u_b(t) = -2u_f e^{-\alpha t}\left(\cos(\gamma t) + \frac{\alpha}{\gamma}\sin(\gamma t)\right) + 2u_f \qquad (3.99)$$

Figure 3.27 shows the capacitor voltage as a function of time. As it can be seen in this figure, at $t = \dfrac{\pi}{\gamma}$, the voltage value at the capacitor can reach as high as four times the input voltage. This voltage can cause a great deal of damage to the system. The abovementioned equations are also valid for the case illustrated in Figure 3.28. In this figure, a three-phase winding is configured as Y connection, which can be seen in high voltage transformers and reactors. It can be seen that we have a lightning strike to one phase of the transmission line. Now, at the neutral point, the stray capacitor C can be formed as a connection from this point to ground. Therefore, we have the connection shown in Figure 3.26. In this condition, the start connection neutral point

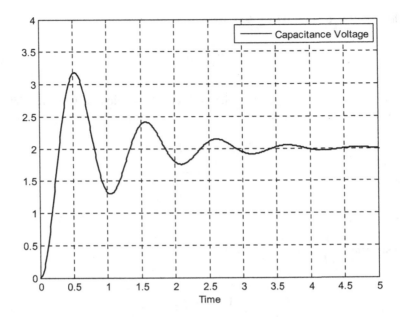

FIGURE 3.27 Voltage at the end of the line with series inductor and capacitor.

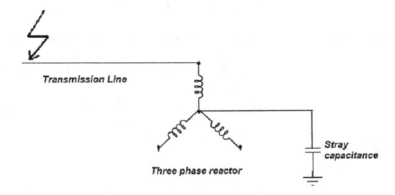

FIGURE 3.28 Three-phase inductor and stray capacitor from neutral to ground.

voltage can reach four times the entering transient voltage wave. In order to prevent the mentioned situation, an arrester will be installed at the neutral point of the star connection, or this point must directly or indirectly be connected to the ground.

3.5.5 COMBINATION OF INDUCTORS AND CAPACITORS IN THE MIDDLE OF THE LINE

One of the ways in fast transmission of data in power systems is the utilization of PLC (power line carrier). The information will be transmitted to the line at high frequencies so that they can easily be isolated at the end of the line. Line traps are used

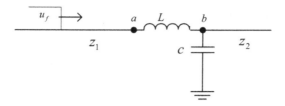

FIGURE 3.29 Capacitor–inductor combination in the middle of the line.

in order to send and receive these signals without loss of energy at power frequency. Line traps are installed on two out of three phases. Line trap is a filter, which can simply be modeled by a combination of an inductor and a capacitor in the middle of one phase of the transmission line as shown in Figure 3.29.

Voltage at a and b can be found as:

$$\begin{cases} u_a = u_{f_1} + u_{r_1} \\ u_b = u_{f_2} \end{cases} \tag{3.100}$$

$$u_a = u_{f_1} + u_{r_1} = L\frac{di}{dt} + u_b \tag{3.101}$$

Considering the figure, we can write:

$$i = i_{f_1} + i_{r_1} = C\frac{du_b}{dt} + i_{f_2} \tag{3.102}$$

Taking the derivative from both sides of the equation, we will have:

$$\frac{di}{dt} = C\frac{d^2u_b}{dt^2} + \frac{di_{f_2}}{dt} \tag{3.103}$$

If we combine Equations (3.101) and (3.103), we will have:

$$u_a = u_b + L\frac{di_{f_2}}{dt} + LC\frac{d^2u_b}{dt^2} \tag{3.104}$$

Considering the fact that $u_a = u_{f1} + u_{r1}$, we can write:

$$u_{f_1} + u_{r_1} = u_b + L\frac{di_{f_2}}{dt} + LC\frac{d^2u_b}{dt^2} \tag{3.105}$$

Multiplying Equation (3.102) by Z_1, we will have:

$$u_{f_1} - u_{r_1} = Z_1 C\frac{du_b}{dt} + \frac{Z_1}{Z_2}u_b \tag{3.106}$$

Adding Equations (3.105) and (3.106) will give:

$$2u_f = LC\frac{d^2u_b}{dt^2} + \left(Z_1C + \frac{L}{Z_2}\right)\frac{du_b}{dt} + \left(1 + \frac{Z_1}{Z_2}\right)u_b \qquad (3.107)$$

With the characteristic equation below:

$$LCm^2 + \left(Z_1C + \frac{L}{Z_2}\right)m + \left(1 + \frac{Z_1}{Z_2}\right) = 0 \qquad (3.108)$$

The roots of this equation are:

$$m = -\alpha \pm \gamma$$

$$\alpha = \frac{Z_1}{2L} + \frac{1}{2CZ_2} \qquad (3.109)$$

$$\gamma = \frac{Z_1^2}{4L^2} + \frac{1}{4C^2Z_2^2} + \frac{1}{LC}\left(\frac{Z_1}{2Z_2} - 1\right)$$

Initial conditions are formed in Equation (3.110).

$$\begin{cases} u_b(0) = 0 \\ i_L(0) = 0 \end{cases} \Rightarrow \frac{du_b(0)}{dt} = 0 \qquad (3.110)$$

Therefore, voltage at b will be:

$$u_b(t) = \frac{A}{2\gamma}[-(\alpha + \gamma)e^{\gamma t} + (\alpha - \gamma)e^{-\gamma t}]e^{-\alpha t} + A \qquad (3.111)$$

Now, based on the parameters given in Equation (3.10), this voltage can be analyzed to estimate possible transient overvoltages and overcurrents, which must be used for line trap design.

3.6 PSCAD EXAMPLES

3.6.1 OVERHEAD LINE AND CABLE CONNECTION

In this subsection, the system in Figure 3.9 is modeled and simulated in PSCAD. The breaker closes at $t = 1$ μs. Figures 3.30 and 3.31 are illustrating the PSCAD representation and system results, respectively.

As it can be seen in Figure 3.31, the voltage wave reaches the end of the line after 1 μs, which is the traveling time for the overhead line. After another 0.5 μs, the wave will reach the end of the cable; this time is also the same as the traveling time in the

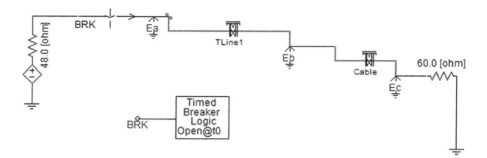

FIGURE 3.30 Under study system.

cable. Voltage magnitude at the sending end of the line is 125 kV and this is exactly the same as the value calculated in Figure 3.10.

3.7 CLOSING BREAKER IN LINE ENERGIZATION

In this section, we are assessing the effect of closing resistance in reducing overvoltages. Figure 3.32 shows a transmission line with the surge impedance of 240 Ω and 1 μs traveling time. The line is energized through a voltage source of 150 kV and 48 Ω internal resistance.

At first, the line is energized directly at $t = 1$ μs. As it is illustrated in Figure 3.33a, voltage at the end of the line reaches 250 kV. In the second condition, a 96 Ω resistance is used for energization. Breaker 1 closes at $t = 1$ μs. Figure 3.33b shows that in this situation, the voltage at the end of the line is less than 200 kV. After the voltage wave reached the stable point, breaker 2 closes to alleviate the losses in the 96 Ω resistance.

3.8 CAPACITOR CONNECTION AT THE END OF THE LINE

The material studied in Section 3.5.1 are modeled in this subsection. Figure 3.34 illustrates the PSCAD implementation for the concepts covered in Section 3.5.1. The system contains a line with the surge impedance of 240 Ω with traveling time of 1 μs. The system source has the voltage magnitude of 150 kV with zero internal resistance. End of the line is terminated with a 1.0 μF capacitor.

It is important to note that in Figure 3.17 and related calculations in Section 3.5.1, it is assumed that the voltage at the sending end of the line is without reflection. In order to model the mentioned situation in PSCAD, the traveling time of the line is considered to be a large number. This will take away the possibility of going back to the sending end from the wave in the period of this study.

Therefore, in this simulation, traveling time is considered to be 1 ms. Results of the simulation are provided in Figure 3.35. As it can be seen, the voltage at the end of the line (i.e., capacitor terminals) will reach the twice of the system sending end voltage (compatible with calculations in Figure 3.18).

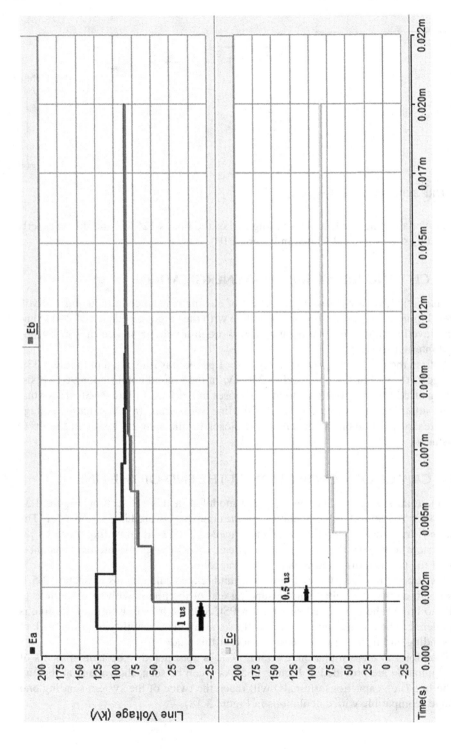

FIGURE 3.31   Simulation results.

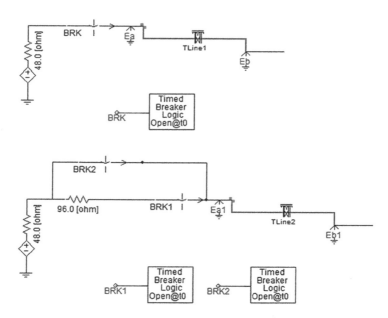

FIGURE 3.32 The under-study system for line energization.

3.9 INDUCTOR CONNECTION AT THE END OF THE LINE

This subsection is illustrating simulations related to Section 3.5.2. Figure 3.36 contains a transmission line with a surge impedance of 240 Ω and traveling time of 1 μs. The system source is a 150 kV voltage source with an internal resistance of zero. End of the line is terminated by a 0.1 H inductor.

Again, traveling time of the transmission line is considered to be 1 ms. Results of the simulation can be seen in Figure 3.37. As it is shown, voltage at the end of the line reaches twice the voltage of the source and will gradually be damped (compatible with calculations in Figure 3.22).

3.10 PARALLEL FILTER LC CONNECTION AT THE END OF THE LINE

In this section, simulations related to Section 3.5.3 have been conducted. Figure 3.38 contains a transmission line with the surge impedance 240 ohm and traveling time of 1 μs. This line is connected to a source of 150 kV with zero internal resistance. End of the line is terminated with a parallel circuit of a 0.1 H inductor and a 1 μF capacitor.

Again, we consider a large traveling time of 1 ms for the line, so the wave does not get a chance to travel back to the sending end of the line in the period of this study. Results of this study can be seen in Figure 3.39. Voltage at the end of the line is compatible with the calculations illustrated in Figure 3.25.

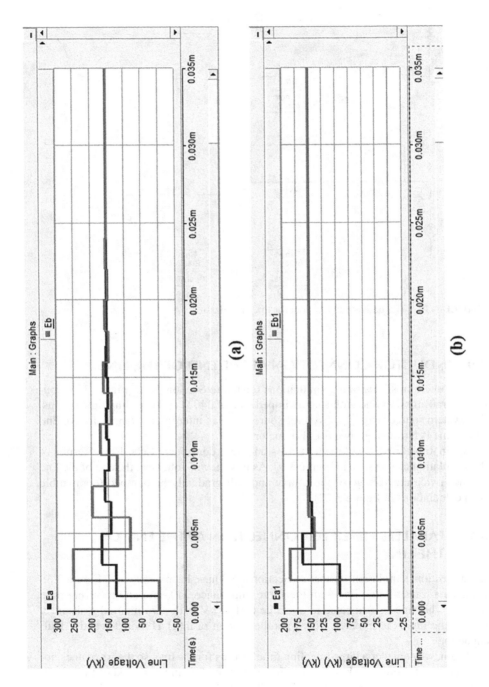

FIGURE 3.33   Simulation results of line energization.

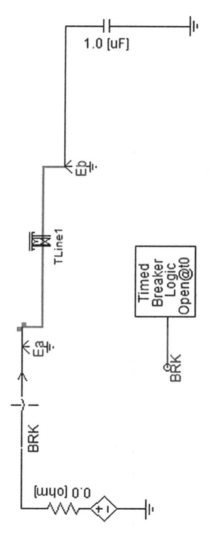

FIGURE 3.34 Capacitor connection at the end of the line.

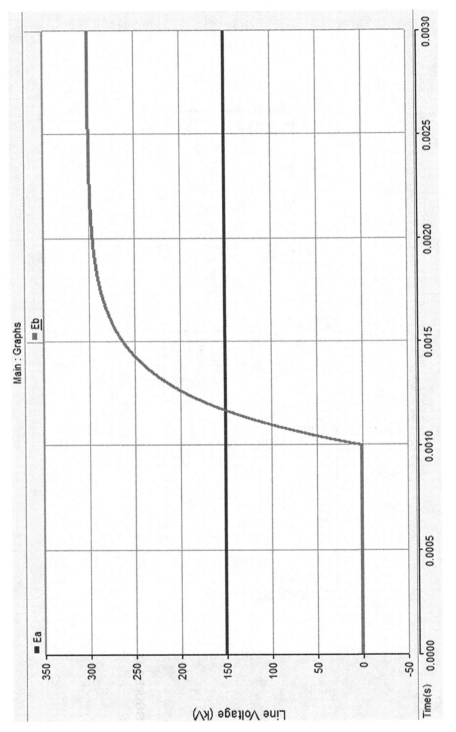

FIGURE 3.35 Simulation results of capacitor connection at the end of the line.

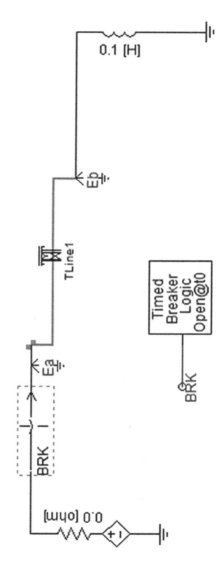

FIGURE 3.36 Inductor connection at the end of the line.

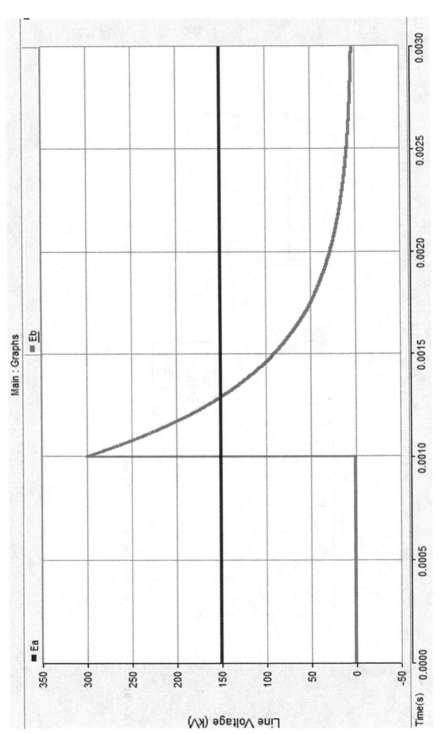

FIGURE 3.37 Simulation results of inductor connection at the end of the line.

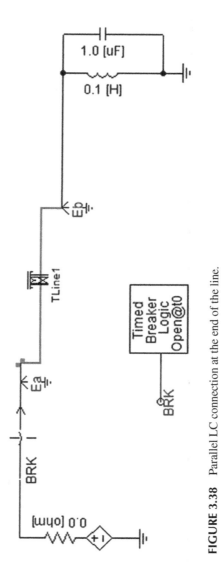

FIGURE 3.38 Parallel LC connection at the end of the line.

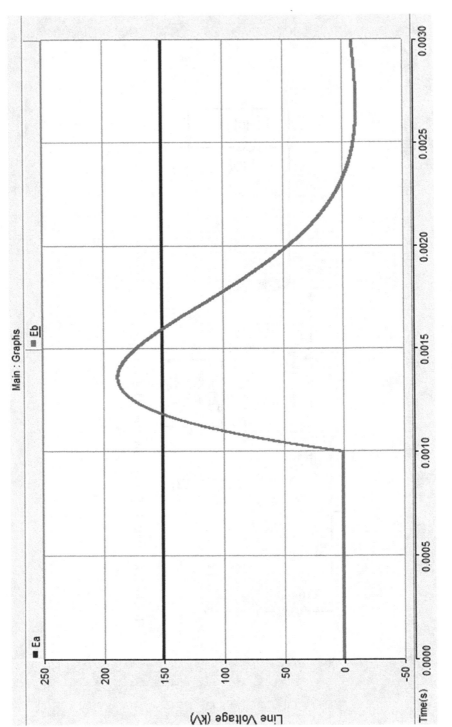

FIGURE 3.39 Simulation results of parallel filter LC connection at the end of the line.

Problems

3.1 Or the power system depicted in the figure, there is a step traveling wave U_{f1} at the beginning of the line with the characteristic impedance Z_1. Write the related differential equation of the voltage at the bus with the reactor L.

3.2 Using lattice diagram on the figure below:
Find voltage at the transformer terminals (i.e., $Z_t = R$) after n reflections. Prove that the magnitude of the voltage at transformer terminals is independent of the cable characteristic impedance (Z_C).
How does the cable affect the system responses during transients?

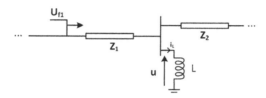

FIGURE P3.1 System diagram.

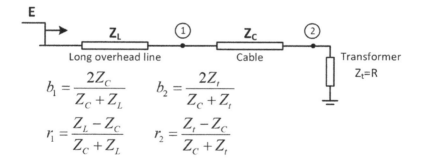

$$b_1 = \frac{2Z_C}{Z_C + Z_L} \qquad b_2 = \frac{2Z_t}{Z_C + Z_t}$$

$$r_1 = \frac{Z_L - Z_C}{Z_C + Z_L} \qquad r_2 = \frac{Z_t - Z_C}{Z_C + Z_t}$$

FIGURE P3.2 System diagram.

BIBLIOGRAPHY

1. Lou van der Sluis, *Transients in Power Systems*, John Wiley & Sons, 2001.
2. Hermann W. Dommel, *EMTP Theory Book*, Microtran Power System Analysis Corporation, 1996.
3. Juan A. Martinez-Velasco, *Transient Analysis of Power Systems: A Practical Approach*, John Wiley & Sons, 2019.
4. Yoshihide Hase, *Handbook of Power System Engineering*, John Wiley & Sons, 2007.
5. Zhengyou He, *Wavelet Analysis and Transient Signal Processing Applications for Power Systems*, John Wiley & Sons, 2016.

6. M. M. Shabestary, A. J. Ghanizadeh, G. B. Gharehpetian, and M. Agha-Mirsalim, "Ladder Network Parameters Determination Considering Non-dominant Resonances of Transformer Winding", *IEEE Transactions on Power Delivery*, Vol. 29, No.1, February 2014, pp. 108–117.

7. M. Mokhtari and G. B. Gharehpetian, "CN Tower Modeling Using EMTP and Taking into Account Frequency Dependent Grounding", *5th Iranian Conference on Engineering Electromagnetics, ICEEM 2017*, 19–20 April 2017, Tehran, Iran.

4 Lightning-Induced Transients

4.1 INTRODUCTION

There are around 2000 thunderstorm per second in the world. This will produce 100 lightning strikes to the earth per second (more than 8 million lightning strikes per day). Lightning causes 100 deaths and 250 injuries per year in the United States. This is quite considerable compared to other climate disasters. In 1746, Benjamin Franklin started studying lightning by using capacitors. He observed many similarities between lightning and the sparks among energized capacitors such as color, smell, and appearance. Before his observations, other research materials suggested that electricity and lightning have the same nature. But this has been proven and Franklin was determined to prove that with a series of experiments. In 1750, Franklin wrote a letter to his friend (P. Collinson) in London, and in that letter, he suggested an experiment in a room. Collinson became his agent in publishing the paper in the *Philosophical Transaction of the Royal Society*. The experiment in the room was done by a sharp rod with the length of 6–9 m. The rod was installed on top of this room. The end of the rod was placed on a conductive pedestal, which was isolated from the ground. In the event of lightning, the observer should stand on the pedestal and hold the rod with one hand. The goal was to see if there will be any sensation in the other hand when the lightning strikes. Franklin had mentioned that the rod needed to be grounded to prevent any hazard to the examiner. Before Franklin in 1752, Dalibard, who was French, had done the same experiment and saw the sparks on his hands and it proved that lightning has electricity. There have been other experiments with similar results. For example, Richman in Russia conducted the same experiment and lost his life when the lightning struck his body. In June 1752, Franklin performed a more elaborated experiment and called it the kite experiment. He argued that if wire conductors are connected to the kite, this combination will act as a longer rod. These wires were connected to a switch and an observer held it with silk that was acting as an insulator. In this experiment, when he attached his ring to the switch, there was a spark. Franklin did not publish the results of the experiment until 1788. The common belief is that Franklin did not know about the results of the room experiment that was conducted by Dalibard before conducting the kite experiment. In 1749, Franklin wrote a paper regarding his invention of rod. This paper was published in 1750. His initial theory was that a sharp rod installed on a rooftop of a house can deplete clouds and will prevent the lightning to hit the house. He had mentioned in a paper published in 1755 that even if the rod does not deplete the electricity, it still prevents the house from getting hit by the lightning. Today, we know that the rod will not deplete the cloud, but conducting the lightning toward the rod helps protecting the house from damage by lightning strike. There were disagreements about the shape of the rod with different suggestions such as sharp rod, flat rod, or a rod with a ball on top of

DOI: 10.1201/9781003255130-4

that. Despite the mentioned disagreement, the common consensus was that the rod was successful in preventing the lightning to hit places. Until the early 19th century, the developments in this area were not very vast, as the rod was performing the function that was expected from it. In the early 19th century, when the electric systems started working, lightning became their biggest problem. Any lightning strike on the electric lines would cause a great deal of damage. Some of the electricity producers came up with the solution of disconnecting electricity in case of bad weather in order to prevent any lightning strikes. In this case, the practice was to ground the electric lines. The challenges led to more research on lightning and its mechanisms.

In studying lightning stroke and its effects, the attention needs to be on where the final step of the wave is going to be applied to or in other words where the wave decides to hit. In this study, we explain some of the mechanisms involved in thunderhead with the goal of developing mathematical models.

There will be electric charge separation in the clouds. This has been shown in Figure 4.1. The bottom part of the cloud carries a negative charge and the top part carries a positive charge. With this the ground under the cloud will get charged/influenced positively. There might be pockets of positive charges in the lower part of the cloud as well. Cloud temperature on average can be −20°C, and the speed of wind can be higher than 160 km/s. The average height of the clouds is between 9 and 12 km. But in some cases, the clouds can be as high as 18 km. In non-mountainous places, the heights of the cloud can be 1.5 km. With the charge separation in the cloud, the potential of the midpoint in the cloud becomes higher and this leads to breaking the air at one point. Due to the break in the air, this will form a path between the negative charge in the cloud and the small positive mass at the bottom of the cloud, or the path can be between the main positive and negative accumulation in the cloud. As time goes, the gradient of voltage on the edge of the cloud increases and will cause breaking the air toward the ground. In this condition, wave will act as a forward moving step toward the ground.

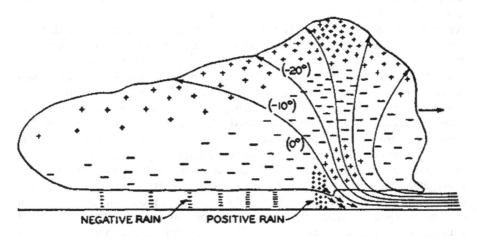

FIGURE 4.1 Electric charge distribution in the cloud.

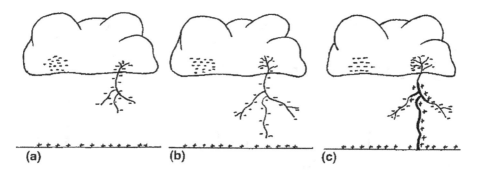

FIGURE 4.2 (a) First strike. Forward moving wave starts at the cloud. (b) Forward moving wave reaches the ground. (c) Upward moving tunnel toward the cloud.

Figure 4.2 shows the general process of lightning strike. The stepped leader travels toward the earth in delaying steps of almost 50 m. After each advancement, there is a short delay. Later, the step will start moving in one or many directions. The step wave traveling time is 50 µs when close to the cloud and 13 µs close to the earth. The speed of the forward traveling wave is low and it is about 0.1% of the speed of the light. The forward moving wave cannot be seen with the naked eyes, and it has a current of 50–200 A.

Stepped leader or the downward moving wave travels toward the earth. In this moment, the back stroke from the earth reaches the downward moving wave. The upward moving wave will travel toward the cloud with the speed of 10–3% of the speed of light. This wave is visible to the naked eye. The current brought to the earth by this upward channel can have the magnitude of 200 kA, but on average, the magnitude is 33 kA. The temperature of this channel is about 28,000°C, which is around five times the temperature of the surface of the sun. The fast increase in temperature produces an impulse wave that is audible. The upward and downward moving waves have the average wavelength of 5–6 km.

As mentioned above, we discussed the mechanism of the first lightning stroke. Each lightning can contain up to 54 strokes, and the average for this number is 3. Figure 4.3 shows the aforementioned structure. After 10–100 ms, the second leader wave, which is called a dart leader, travels from the cloud to earth. With the advent of dart, another part of the electric charge gets depleted. The dart leader, as it is suggested by its name, does not have steps and travels directly to ground (Figure 4.3b). In order for the mentioned leader to start traveling, some of the charge in the cloud has to get depleted. The dart leader does not have steps and moves directly to ground. Its speed is around 1% of the speed of light, which is 10 times higher than the speed of the leader stepped wave. The reason for higher speed and direct stroke of the dart wave is that the first wave (forward traveling wave) has ionized the atmosphere and has provided a path for the dart wave. When the dart's head reaches the earth, an upward channel starts from the earth. The wave traveling in the upward direction will hit the downward traveling wave. An electric current is discharged to the earth with the magnitude of 40% of the first stroke. The residual charges in the cloud can produce more dart leaders.

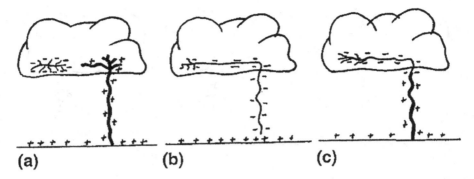

FIGURE 4.3 The second stroke. (a) Upward traveling canal reaches the cloud. (b) Dart leader travels toward ground. (c) The upward traveling canal will be formed.

Let us examine the stepped or the downward leader as shown in Figure 4.4. This will help us understand the mechanism of first stroke. As it can be seen in Figure 4.4, the downward leader includes two parts. There is a highly conducting core or channel and a negative space charge, which is in front of the core and surrounds the core. The conducting channel can have the diameter of 2 mm with the voltage gradient of 50 kV/m. The electric charge of the cloud will reduce as the leader travels. The residual charge in the cloud will be distrusted in the air as corona. Figure 4.4 illustrates the downward moving leader in the instant of stepping process. In this situation, the space charge is maximized and the leader wave will do its second step. The potential of the leader wave is around 50,000 kV as it is shown. Fast extension of the channel continues up to the border of the corona sheath. At this time, the leader stops, while corona will get propagated around it and later step happens again. This process happens from the cloud (where there are 50 μs steps) to earth (where there are 13 μs steps). As an approximation, as the wave approaches ground, it will maintain a constant speed.

The last stage of the stroke where the downward leader approaches the earth is illustrated in Figure 4.5. As the final step happens, the wave becomes the high speed, high current return stroke. The downward moving leader has the speed of 30 cm/μs if the voltage gradient of 50,000 kV at the leader wave. In this figure, the leader wave is approaching a 30 m tall building. Part A of the figure shows that the corona discharge with positive polarity gets started at the tip of the building. The distance between points a and b, which are the top of the building and the tip of the downward leader, respectively, is 82.5 m. In part B of the figure, the leader has traveled downward another 1.5 m and approaches the corona appeared at the top of the building. At this limiting point, which we consider as $t = 0$, the downward leader decides to strike the top of the building and not the earth. Some laboratory measurements on rod-rod gaps show the breakdown gradient for negative polarity is 605 kV/m. Remember that the 50,000 kV voltage gradient of the leader wave and dividing it by the downward gradient (i.e., 605 kV/m), will give us the distance between the points a and b in part B of Figure 4.5. This distance is 82.64 m. In this distance, there will be the advent of

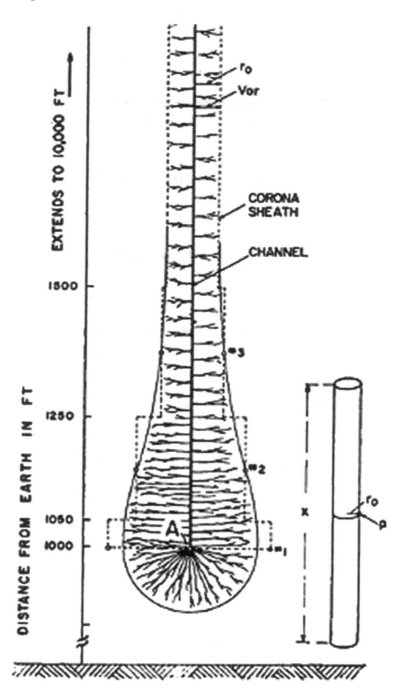

FIGURE 4.4   Downward traveling wave.

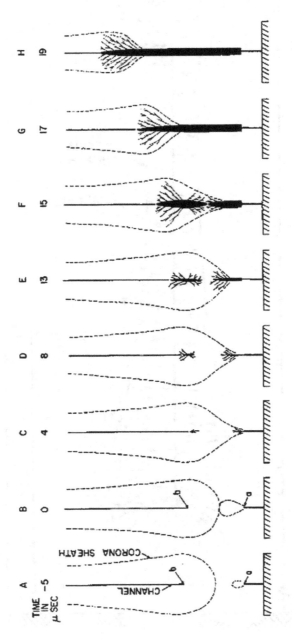

FIGURE 4.5 Upward channel development.

the last step and it is called a "striking step." Upward and downward moving channels continue their directional movement as can be seen in part C of Figure 4.5. The speed of these channels approaching each other is very high, and therefore, it produces increasing current. In part F of the figure, the channels reach each other and the electric current peaks. In G and H parts, there is only one channel and it is upward moving. The charge that has been deposed previously by the downward leader is tapped by the formed channel and the charge gets drained to earth. The corona sheath of the downward leader produces electric currents and not the cloud charge.

The above paragraph explained the "last step of the first stroke" with a negative polarity "downward flash." This is a very common type of flash. There are three other types, which we try to briefly explain in here.

- The first type is the "negative downward flash" hitting buildings shorter than 100 m with 33 kA average current.
- The second "negative upward flash" hits taller buildings. In areas with heights of 650 m from sea level, they hit buildings around 80 m tall. The average of current is 25 kA.
- The third type is the "positive upward flash" with the current magnitude of 1.2–2.2 times the "negative downward flash." Positive flashes usually have one stroke per flash, which happens at the start or finish of the storm in oceans. They are more prevalent in the winter season. Usually, 2–10% of the flashes are from this type.

Around 85–95% of the lightning strokes happen on the buildings with the height 100 m from ground are "negative downward waves." Around 5–15% are for the "negative or positive upward." Except for mountainous areas or places with long rivers, negative downward moving wave is concerning for transmission lines or high voltage substations. These types and associated numbers are accumulated from the real cases of lightning stroke that have happened around the world. Data regarding "positive downward flash" (i.e., fourth type) is not available.

4.2   GENERAL CHARACTERISTICS OF LIGHTNING SURGES

Lightning is the electric discharge of the cloud to earth or the discharge from earth to cloud. Also, lightning can happen because of the electric discharge between two clouds, which is illustrated in Figure 4.6. As mentioned before, this random process has a negative polarity more than 90% of the time. In one lightning stroke, there appears several strikes. In our studies here, lightning is always considered as it is depicted in Figure 4.7.

It is worth noting that the positive polarity lightning waves have larger amplitudes. Figure 4.7 shows the standard wave function for lightning. The mathematical model is as follows:

$$i(t) = \alpha_k I_m (e^{-\alpha t} - e^{-\beta t}) \tag{4.1a}$$

FIGURE 4.6   Lightning flashes.

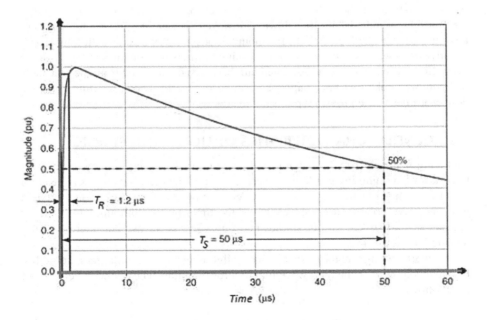

FIGURE 4.7   Impulse function.

$$i(t) = \alpha_k I_m \left(e^{\frac{-t}{T_s}} - e^{\frac{-t}{T_R}} \right) \tag{4.1b}$$

In the above equation, I_m is the magnitude of the first wave and it is 31 kA. T_R is the rise time, which is 1.2 μs. T_S is the fall time and that is around 50 μs.

IEEE standard suggests using Equation (4.2) for calculating the probability of having the lightning stroke current greater than I:

$$P(i > I) = \frac{1}{1 + \left(\dfrac{I[kA]}{31} \right)^{2/6}} \tag{4.2}$$

The amplitude of the lightning current is variable. Measurements show that I_m can vary from 1 to 220 kA. But the probability of having the amplitude between 5 and 100 kA is higher. Lightning stochastically analysis offers the amplitude of 20 kA for lightning current.

4.3 TYPES OF STROKE

4.3.1 DIRECT STROKE TO PHASE CONDUCTOR

In this case, lightning hits one of the phases in the transmission line. In order to analyze this condition, we model the lightning as an ideal current source, which hits a line with characteristic impedance of Z. As the lightning current goes through the line, the voltage wave components of u_f and u_r will be produced. After lightning stroke to the conductor, two waves will appear, one traveling to the right and the other traveling to the left. Figure 4.8 illustrates the aforementioned waves. No matter wherein the lightning hits, we can assume that the impedance will be the same characteristic impedance. Figure 4.9 shows the equivalent circuit that has been made considering what is mentioned here.

According to Figure 4.9, we will have the potential at point a as:

$$u_a = u_f = u_r = \frac{Z}{2} I \tag{4.3}$$

Let us consider the following numerical example [i.e., (4.4)].

$$\begin{cases} Z = Z_c = 300\Omega \\ I = I_m = 30kA \end{cases} \Rightarrow u_f = \frac{1}{2} \times 300 \times 30 \times 10^3 = 4.5MV \tag{4.4}$$

This voltage value cannot be tolerated by the phase conductor and it causes an electric discharge between the conductor and tower. This phenomenon is called "flashover."

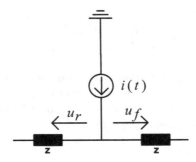

FIGURE 4.8 Current source model for lightning.

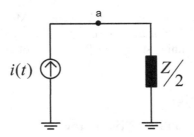

FIGURE 4.9 Equivalent circuit.

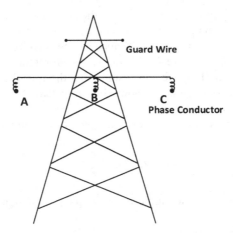

FIGURE 4.10 Phase conductor and guard wire arrangement in transmission line.

4.3.2 GROUND CONDUCTOR OR TOWER STROKE

Figure 4.10 illustrates the arrangement of three phase lines and their ground conductor. Figure 4.11 shows the protected area that the ground wire makes for the transmission lines. The main responsibility of the guard wire is to protect the transmission line conductors from being exposed to direct lightning strokes.

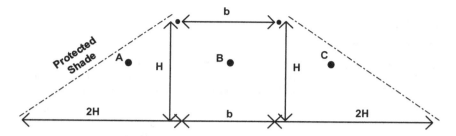

FIGURE 4.11 Geometry of line conductors and guard wires and protected area/angle.

The characteristic impedance of the guard wire is much smaller than the characteristic impedance of the phase conductors. Now let us consider the numerical example in Equation (4.4) and calculate the overvoltage for the ground conductor as in Equation (4.5).

$$\begin{cases} Z = Z_c = 14\Omega \\ I = I_m = 30kA \end{cases} \Rightarrow u_f = \frac{1}{2} \times 14 \times 30 \times 10^3 = 210kV \tag{4.5}$$

From Equation (4.5), we can conclude that if the guard wire characteristic impedance is designed to be low, then the overvoltage resulting from lightning strike will decrease. Power delivery limitations are irrelevant in designing guard wires. Therefore, it is possible to choose the characteristic impedance as small as possible.

Characteristic impedance (Z_c) value depends on the soil resistivity of the area where the tower is installed. In Equation (4.5), we showed that the tower will tolerate a voltage of around 210 kV. Now, we need to consider the "Back Flashover" that can appear from the tower to the phase conductor. This has been shown in Figure 4.12. It is worth noting that the guard wire is connected to the tower and the tower is grounded

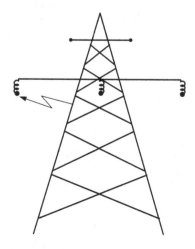

FIGURE 4.12 Back flashover.

TABLE 4.1

Transient Overvoltages

System Nominal Voltage	Overvoltage Due to Lightning Strike on the Tower or the Guard Wire (kV) (4–6 with the Assumption of 210 kV Overvoltage)	Overvoltages Due to Switching (kV)	Priority Consideration for the Voltage Level
400	775.69	1131.37	Switching
230	535.27	650.54	Switching
132	396.68	373.35	Lightning
63	299.10	178.19	Lightning

(connected to the ground). We can say that the maximum voltage difference between the guard wire and phase conductor can be calculated as in Equation (4.6).

$$\sqrt{2} \times V_{LL-RMS} + 210[kV] \tag{4.6}$$

The above calculation has been done in the worst-case scenario of having the phase voltage at its peak. In Chapter 3, we examined the fact that in case of open circuit line, the overvoltage on that line can be as high as twice the amplitude of the traveling wave. Therefore, the worst-case scenario calculation will give us $2\sqrt{2}\, V_{LL-RMS}$. Let us compare the overvoltage because of switching it due to lightning. Table 4.1 summarizes the mentioned comparison.

As it can be observed in Table 4.1, for instance, the overvoltages due to lightning are more important than the overvoltages due to switching for voltages below 132 kV or subtransmission systems.

4.3.3 NEARBY STROKES

If the lightning stroke happens on a tree or the ground that is near a transmission line (up to 200 m), then electromagnetic rules state that there will be inductions on the line, although it has not been exposed directly to the lightning. The following are the types of inductions that can happen:

1. Electrostatic induction (influence)
2. Magnetic induction (induction)

In the second type, the change in magnetic flux $\left(\dfrac{d\varphi}{dt}\right)$ is the reason for induction. This phenomenon usually happens in distribution grids. The reason for that is that the trees and buildings are taller than the installed distribution lines; as a result, the possibility of the lightning hitting the trees or buildings is higher.

4.4 SYSTEM RESPONSE TO VOLTAGE ARBITRARY PULSES

In order to analyze the response of the system to the imposed arbitrary voltage wave, we need to have the response to the step wave available. Let us assume that $u(t)$ is the system step response; then, using convolution integral, $f(t)$ is the system response to the imposed voltage, which is shown as follows:

$$f(t) = U(t) \times u(0) + \int_{\xi=0}^{t} U(t-\xi) \frac{\partial u(\xi)}{\partial \xi} d\xi \qquad (4.7)$$

Figure 4.13 shows a line that is terminated to a capacitor bank. In Chapter 3, we analyzed how traveling wave will affect the voltages in this system. Using the method mentioned in that chapter, we can find the voltage appearing across the capacitor. In this figure, U^* is the magnitude of the step wave.

We assume a wave similar to what is shown in Figure 4.14 travels through a line that is terminated with a capacitor bank. The imposed voltage of a lightning, as Figure 4.7 suggests, can be formulated as Equation (4.8):

$$U(t) = U^*(e^{-\alpha t} - e^{-\beta t}) \qquad (4.8)$$

The considerable point in Figure 4.14 is the fact that first the rising wave will appear in the line and starts traveling through it. Therefore, the waveform shown in this figure is similar to the one presented in Figure 4.7.

In Chapter 3, we found that the voltage across the capacitor can be written as follows:

$$u(t) = 2U^* \left(1 - e^{\frac{-t}{CZ}} \right) \qquad (4.9)$$

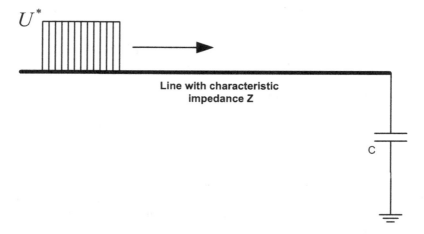

FIGURE 4.13   Step wave traveling through a transmission line.

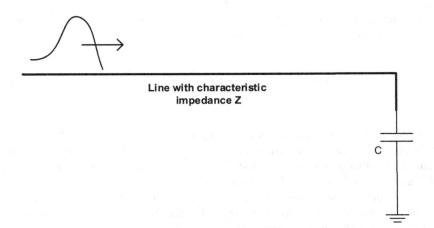

FIGURE 4.14   Arbitrary wave traveling through a transmission line.

If we take a derivative of (4.9) and replace t with ξ, we will have:

$$\frac{\partial u(\xi)}{\partial \xi} = \frac{2U^*}{CZ} e^{\frac{-\xi}{cz}} \tag{4.10}$$

As we discussed before, the capacitor will act as a short circuit at the instant that the wave approaches the capacitance ($u(0) = 0$). If we consider $\tau = ZC$, then Equation (4.7) will be changed to Equation (4.11)

$$f(t) = \frac{2U^{*2}}{\tau} \int_{\xi=0}^{t} (e^{-\alpha(t-\xi)} - e^{-\beta(t-\xi)}) e^{\frac{-\xi}{\tau}} d\xi \tag{4.11}$$

Integrating Equation (4.11) and placing the limits there, we will have:

$$f(t) = \frac{2U^{*2} e^{-\alpha t}}{\alpha\tau - 1}\left[e^{\left(\alpha t - \frac{t}{\tau}\right)} - 1\right] - \frac{2U^{*2} e^{-\beta t}}{\beta\tau - 1}\left[e^{\left(\beta t - \frac{t}{\tau}\right)} - 1\right] \tag{4.12}$$

After simplification, we get Equation (4.13).

$$f(t) = \frac{2U^{*2}}{\alpha\tau - 1}\left[e^{-\frac{t}{\tau}} - e^{-\alpha t}\right] - \frac{2U^{*2}}{\beta\tau - 1}\left[e^{-\frac{t}{\tau}} - e^{-\beta t}\right] \tag{4.13}$$

Figure 4.15 shows the voltage that appears across the capacitor in case of step voltage or lightning impulse. We assumed the following values $\alpha = 1/T_S = 1/1.2$ μs and $\beta = 1/T_R = 1/50$ μs. It is obvious that in this figure, the initial instants are the same for both curves, because at the instant that the wave approaches the

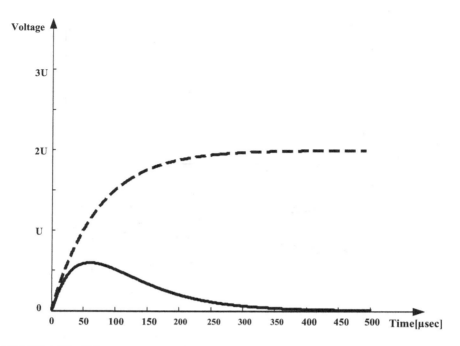

FIGURE 4.15 Voltage across capacitor bank. Step voltage input (dashed line) and lightning impulse (connected line).

capacitance, it behaves like an open circuit, but after this period, their behaviors are different.

4.5   APPLICATION OF LATTICE DIAGRAM

In this part, we analyze the effect of lightning impulse on a transmission line using lattice diagram. The general equation of the impulse wave is repeated here:

$$E(t) = E(e^{-\alpha t} - e^{-\beta t}) \qquad (4.14)$$

Equation (4.14) includes two parts, a short rising wave and a longer falling wave. As we observe, the mathematical functions for both waves are the same. Therefore, we can use one of the two in conducting the analysis. If we start with the rising wave, then the response will be the same for the falling wave by replacing the parameter α with β and multiplying the equation by a negative sign. Figure 4.16 shows the lattice diagram that is used to find the voltage at point 2. Assume it takes T second for the wave to travel from point 1 to point 2. In that time, the point 2 voltage can be calculated as Equation (4.15).

$$V_{21}(t) = E b_1 b_2 e^{-\alpha(t-T)} u(t-T) \qquad (4.15)$$

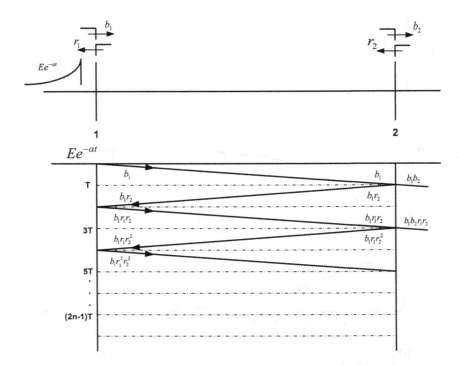

FIGURE 4.16   Lattice diagram lightning stroke on the transmission line.

In Equation (4.15), V_{21} is the voltage at point 2 in the first stroke. $u(t - T)$ is a step function. In the second stroke, the voltage at point 2 will be:

$$V_{22}(t) = E b_1 b_2 r_1 r_2 e^{-\alpha(t-3T)} u(t - 3T) \tag{4.16}$$

Continuing the process in the n-th stroke, the voltage equation is written as:

$$V_{2n}(t) = E b_1 b_2 (r_1 r_2)^{n-1} e^{-\alpha(t-(2n-1)T)} u(t - (2n - 1)T) \tag{4.17}$$

If we assume $Q = b_1 b_2$ and $= r_1 r_2$, we can write the voltage at point 2 after n-th stroke as:

$$V_{2n}(t) = E Q[e^{-\alpha(t-T)} u(t - T) + M \times e^{-\alpha(t-3T)} u(t - 3T) + \cdots \atop + M^{n-1} e^{-\alpha(t-(2n-1)T)} u(t - (2n - 1)T)] \tag{4.18}$$

At $t = (2n - 1)T$, we will have:

$$V_{2n}(t)\big|_{t=(2n-1)T} = E Q[e^{-\alpha 2(n-1)T} + M e^{-\alpha 2(n-2)T} + \cdots + M^{n-2} e^{-\alpha 2T} + M^{n-1}] \tag{4.19}$$

Using the geometric progression given in Equation (4.19), we can simplify this equation in form of (4.20). Again, Equation (4.20) is the sum of voltages at point 2 after n strokes.

$$V_{2n} = EQM^{n-1} \frac{1-(M^{-1}e^{-2\alpha T})^n}{1-M^{-1}e^{-2\alpha T}} = \frac{EQ(M^n - e^{-2\alpha nT})}{M - e^{-2\alpha T}} \qquad (4.20)$$

Observing the fact that the voltage at point 2 is a function of number of strokes (n), it makes sense to calculate the number of strokes, which makes the maximum voltage at 2 be calculated along with the n value that causes the maximum voltage. So, assuming a continuous function, we need to take the derivative of V_{2n} with respect to n and set the derivative to zero as shown in Equation (4.21).

$$\frac{\partial V_{2n}}{\partial n} = 0 \Rightarrow M^n lnM + 2\alpha Te^{-2\alpha nT} = 0 \qquad (4.21)$$

Solution of (4.21) will provide:

$$n_{max} = \frac{ln(2\alpha T) - ln(-lnM)}{lnM + 2\alpha T} \qquad (4.22)$$

In electrical engineering, it is important to investigate any mathematical solutions to make sure the obtained numbers are in normal ranges. Therefore, we consider that the operational boundaries of the system are crucial in accepting the calculated numbers. Special case analysis can highlight the operational boundaries of the system. As it can be seen in these equations, if we set $\alpha = 0$, the wave will be altered to a step wave. The special condition will yield the result of $n = \infty$, which complies with our findings in Chapter 3. This will inform us that the maximum voltage at point 2 happens after a large number of strokes as it is given in Equation (4.23).

$$\lim_{\alpha \to 0} n_{max} = \lim_{\alpha \to 0} \frac{ln(2\alpha T) - ln(-lnM)}{lnM + 2\alpha T} = \infty \qquad (4.23)$$

If we replace Equation (4.22) in Equation (4.21), the maximum voltage at point 2 will be calculated as:

$$V_{2n\,max} = \frac{EQ(M^{n_{max}} - e^{2\alpha n_{max}T})}{M - e^{2\alpha T}} \qquad (4.24)$$

Equation (4.24) will tell us that the maximum voltage at point 2 is a function of T, M, and α. Parameter T will be identified by the length of the line. Parameter $M = r_1 r_2$ is obtained from the line characteristic impedance and α is obtained from the input wave. Therefore, the maximum voltage is a function of the line length, characteristic impedance, and the input voltage. Figure 4.17 shows a diagram of the voltage at the

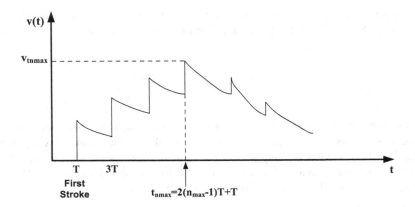

FIGURE 4.17 Voltage at the end of an open circuit line.

end of the open circuit line. This diagram has been constructed using an arbitrary wave coming to the beginning of the open circuit line and applying Equation (4.20).

In Equation (4.14), the rising and falling waves are both exponential functions. Therefore, we can write the same equations for the falling wave and find the sum of the voltages at point 2 as given in Equation (4.25)

$$V_{tn} = \frac{-EQ(M^n - e^{-2\beta nT})}{M - e^{-2\beta T}} \tag{4.25}$$

Using superposition, now, the total voltage at point 2 will be as given in Equation (4.26)

$$V_{tn} = EQ\frac{(M^n - e^{-2\alpha nT})(M - e^{-2\beta T}) - (M^n - e^{-2\beta nT})(M - e^{-2\alpha T})}{(M - e^{-2\beta T})(M - e^{-2\alpha T})} \tag{4.26}$$

Doing a little algebra on the final equation, we can simplify the equation as follows:

$$V_{tn} = EQ\left[\frac{-M(e^{-2\alpha nT} - e^{-2\beta nT}) + M^n(e^{-2\alpha T} - e^{-2\beta T}) + e^{-2(\alpha n + \beta)T} - e^{-2(\alpha + \beta n)T}}{(M - e^{-2\alpha T})(M - e^{-2\beta T})}\right] \tag{4.28}$$

Now, we suggest the reader to drive the equation for the n-th stroke when $n \to \infty$. Also, examine maximizing Equation (4.28).

4.6 TOWER MODELING FOR TRANSIENT STUDIES

In order to analyze the lightning strike on a transmission line, it is required to model the towers in transient conditions. Also, it is required to have a circuit model of transmission towers in order to study the transients in software platforms such as EMTP

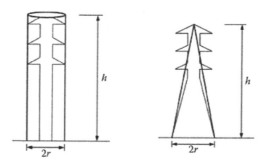

FIGURE 4.18 Cone (right) and cylindrical (left) towers.

or PSCAD. The most common transient model for towers is the transmission line model with a known characteristic wave impedance and a travel time of wave in the length (height) of the tower.

Considering the tower as a cylinder or a cone as shown in Figure 4.18, the characteristic impedance of the line can be calculated based on the height of the tower and the radius of the base of the tower. In a cylindrical condition, the characteristic impedance will be presented as Equation 4.29. In the case of having the tower in the shape of a cone, the characteristic impedance of the line will be presented as given in Equation 4.30.

$$Z = 60\left(\ln\left(2\sqrt{2}\,\frac{h}{r} \right) - 1 \right) \tag{4.29}$$

$$Z = 60\left(\ln\sqrt{2}\,\sqrt{\left(\frac{h}{r}\right)^2 + 1} \right) \tag{4.30}$$

For towers with middle sections or as they call them towers with waist (shown in Figure 4.19), Equation 4.31 will be used to calculate the line characteristic impedance.

$$Z = 60 \ln\left(\cot\frac{\theta}{2} \right) \tag{4.31}$$

In Equation 4.31, we have:

$$\begin{cases} \theta = \tan^{-1}\left(\dfrac{r_{av}}{h} \right) \\[2mm] r_{av} = \dfrac{r_1 h_2 + r_2 h + r_3 h_1}{h} \\[2mm] h = h_1 + h_2 \end{cases} \tag{4.32}$$

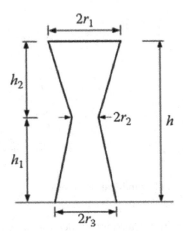

FIGURE 4.19 Waist tower.

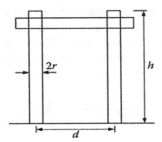

FIGURE 4.20 H-type tower.

Another form of tower is the H-type tower (Figure 4.20) and the characteristic impedance of the transmission line on these towers will be calculated using Equations 4.33 and 4.34.

$$Z = \frac{Z_1 Z_2}{Z_1 + Z_2} \tag{4.33}$$

$$\begin{cases} Z_1 = 60\left(\ln\left(2\sqrt{2}\,\frac{h}{r} \right) - 1 \right) \\[2mm] Z_2 = \dfrac{d\,60\ln\left(2\,\dfrac{h}{r} \right) + hZ_1}{h+d} \end{cases} \tag{4.34}$$

A multiconductor model is also used for the modeling of towers. In this model, the tower is divided into multiple sections. These sections are shown in Figure 4.21.

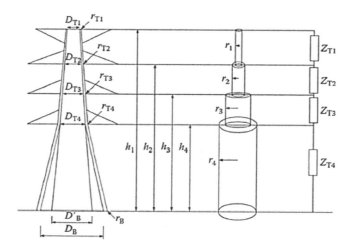

FIGURE 4.21   Tower with a muticonductor model.

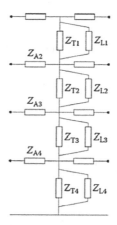

FIGURE 4.22   Equivalent multiconductor model of tower.

An equivalent characteristic impedance will be considered for each section. Finally, the modeling sketch can be seen as illustrated in Figure 4.22.

In this model, Z_T, associated with each part, can be calculated as in Equation (4.35).

$$Z_T = 60\left(\ln\left(2\sqrt{2}\,\frac{h_k}{r_{ek}}\right) - 2\right) \tag{4.35}$$

where

$$r_{ek} = 2^{1/8}\left(\sqrt[3]{r_{Tk}r_B^2}\right)^{1/4}\left(\sqrt[3]{D_{Tk}D_B^2}\right)^{3/4} \tag{4.36}$$

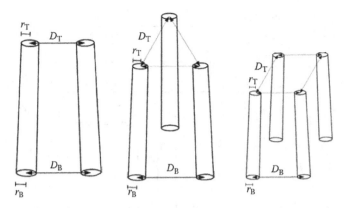

FIGURE 4.23 Schematic to obtain equivalent radius for each section.

The parameters in 4.36 are shown in Figure 4.23, based on the considered conductors for each part.

Usually, $Z_L = 9Z_{Tk}$, which is the bracing effect of the tower. In this model, the tower crossarm is considered as Z_{Ak} as presented in Equation 4.37.

$$Z_{AK} = 60 \ln \frac{2h_{AK}}{r_{AK}} \tag{4.37}$$

where h_{Ak} is the height and r_{Ak} is the equivalent crossarm radius.

Multistory transmission tower model is another common way of modeling towers. This model considers several parts for the tower and each part has a transmission line model with a characteristic impedance in series with a parallel connection of resistance and inductance circuit. Figure 4.24 shows the details of this type of modeling.

In this model, the characteristic impedance of each part is obtained using experimental cases or measurements. Resistance and inductance in each part can be calculated using the following equations.

$$R_i = \frac{-2Z_{T1} \ln \sqrt{\gamma}}{h_1 + h_2 + h_3} h_i \tag{4.38}$$

$$R_4 = -2Z_{T2} \ln \sqrt{\gamma} \tag{4.39}$$

$$L_i = \alpha R_i \frac{2h}{c} \tag{4.40}$$

In equations above, γ is the propagation factor and it is in the range of 0.7 to 0.8. α is the damping factor and it is in the range of 0.8 to 1.0.

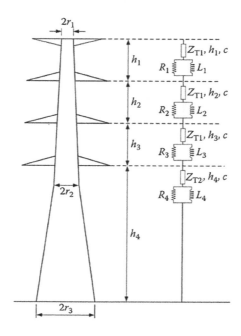

FIGURE 4.24 Multistory tower.

4.7 PSCAD EXAMPLES

In this section, a lightning wave has been modeled in PSCAD. The wave is constructed using Equation 4.1b. This wave is imposed on a transmission line that is terminated with a capacitor of 0.3 μF. The lightning wave has the maximum of 150 kV with the rise time 1.2 μs and the fall time of 50 μs. Figure 4.25 shows the transmission

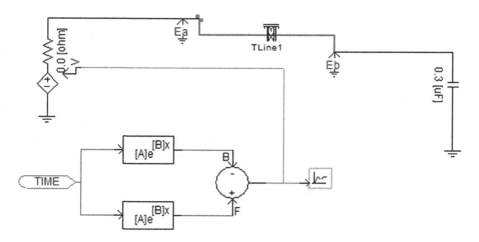

FIGURE 4.25 Simulation system with lightning impulse.

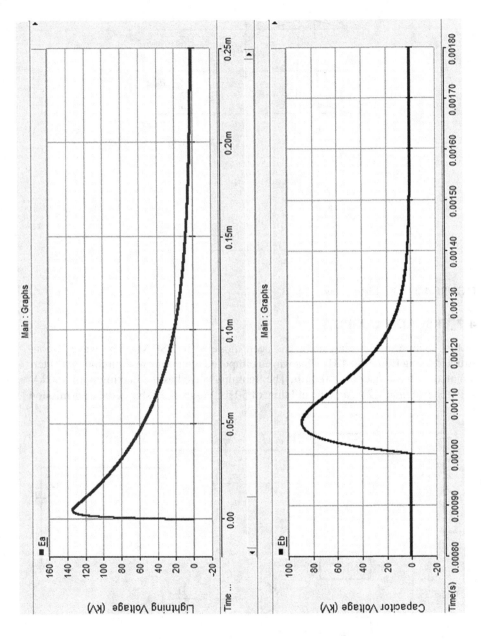

FIGURE 4.26 Lightning and capacitor voltages.

line with the capacitor and also the implementation of the lightning pulse in PSCAD. According to Figures 4.13 and 4.14 and also the calculations presented in subsection 4.4, it is assumed that the voltage does not reflect at the sending end of the line. In order to consider the explained situation in PSCAD modeling (i.e., no reflection at the sending end), the transmission line traveling time is assumed to be really large. With that during the simulation time, the traveling wave does not get a chance to get back to the sending end of the line. In simulations, 1 ms traveling time has been considered.

In Figure 4.26, the plots of the lightning and line voltage are illustrated. As it can be seen that after 1 ms, the wave will reach the end of the line.

Depending on the value of the capacitor bank, the voltage peak will be different at both sides of the line.

Problems

4.1 Assume a transmission tower can be modeled by a series RL circuit and has a ground wire with the characteristic impedance of $Z(\Omega)$. If a lightning surge with the function $i(t) = I_{max}(e^{-\alpha t} - e^{-\beta t})$ hits the ground wire in the vicinity of the tower, find the voltage that is induced across towers, ignoring reflections in adjacent towers.

4.2 A lightning hits the transmission line with the characteristic impedance of $Z_c = 300\,\Omega$, as shown in the figure. If the lightning has the following time function $i(t) = I_{max}(e^{-\alpha t} - e^{-\beta t})$, find:

a. Voltage at bus 2 when the lightning wave reaches this bus.

b. Specify if the place where lightning hits the line affects the voltage at bus 2. Explain why.

c. Assume the lightning hits directly bus 2; in this case, find the voltage at bus 2 and compare this with the findings in part a.

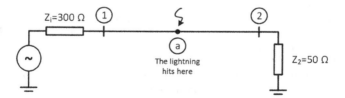

FIGURE P4.2 Circuit diagram.

BIBLIOGRAPHY

1. Andrew R. Hileman, *Insulation Coordination for Power Systems*, CRC Press, 1999.
2. Lou van der Sluis, *Transients in Power Systems*, John Wiley & Sons, 2001.
3. Juan A. Martinez-Velasco, *Transient Analysis of Power Systems: A Practical Approach*, John Wiley & Sons, 2019.
4. Neville Watson and Jos Arrillaga, Power Systems Electromagnetic Transients Simulation, The Institution of Engineering and Technology, 2007.
5. J. C. Das, Transients in Electrical Systems: Analysis, Recognition, and Mitigation, McGraw Hill, 2010.
6. M. Nafar, G. B. Gharehpetian, and T. Niknam, "A New Parameter Estimation Algorithm for Metal Oxide Surge Arrester", Electric Power Components and Systems, Vol. 39, No. 7, July 2011, pp. 696–712 (ISI-ranked).
7. H. Radmanesh, G. B. Gharehpetian, and S. H. Fathi "Ferroresonance of Power Transformers Considering Nonlinear Core Losses and Metal Oxide Surge Arrester Effects", Electric Power Components and Systems, Vol. 40, No. 5, May 2012, pp. 463–479 (ISI-ranked)
8. M. Nafar, G. B. Gharehpetian, and T. Niknam, "Comparison of Parameter Estimation Methods of Surge Arresters Using Modified Particle Swarm Optimization Algorithm", *European Transactions on Electrical Power*, Vol. 22, No. 8, November 2012, pp. 1146–1160 (ISI-ranked).
9. M. Nafar, G. B. Gharehpetian, and T. Niknam, "A New Method for Parameter Estimation of Different Surge Arrester Models", *26-th International Power System Conference*, 31 October–2 November 2011, Tehran, Iran (in Persian).
10. G. B. Gharehpetian, H. Mohseni, and K. Lux, "Overvoltage Studies of Transformer Windings Using Monte Carlo Simulation of Lightning", *Journal of Technical Faculty of Tehran University*, 55, 1995, pp. 73–86.
11. H.S. Tabari and G. B. Gharehpetian, "Suggestion of a Proper Probability Distribution Function for Lightning", *3-th Conference on Probability and Stochastic Process*, 29–30 August 2001 (in Persian).
12. M.H. Nazemi, G. B. Gharehpetion, and H. Javadi, "Effect of High Voltage Cable Characteristics on Damping of Transient Overvoltages due to lightning in Combination of Line, Cable and Transformer", *17-th Power System Conference*, 28–30 October 2002, Tehran, Iran (in Persian).
13. F. Shahniaand and G. B. Gharehpetian, "Lightning Transient Studies of Traction Systems", *14-th Iranian Conference on Electrical Engineering*, 16–18 May 2006, Tehran, Iran (in Persian).
14. F. Shahnia and G. B. Gharehpetian, "Lightning and Switching Transient Overvoltages in Power Distribution Systems Feeding DC Electrified Railways", *3-rd International Conference on Technical and Physical Problems in Power Engineering (TPE-2006)*, 29–31 May 2006, Ankara, Turkey.

5 Energization Overvoltages

This chapter studies the switching-induced transients due to closing a circuit or an apparatus. In the previous chapters, transient main equations are assessed by applying general assumptions. In this chapter, similar to the previous one, we will model and study the transients caused by energization, applying the experimental and operational experiences.

5.1 SOURCE-TYPE EFFECT

We consider the following two conditions in order to study the effect of the source in energizing a transmission line:

a. The source impedance will be ignored.
b. The source will be modeled as a unit step function.

Remembering the fact that transients happen in very small time scales such as microsecond and less than 1 ms, therefore, using the step function as the input source will not cause error in the study. Even a sine wave with 50 or 60 Hz frequency will not have a remarkable change in magnitude considering the time scale of less than 1 ms. Figure 5.1 shows the lattice diagram of an open circuit line energized by a step function. Figure 5.2 shows the voltage diagram of the sending and receiving ends of

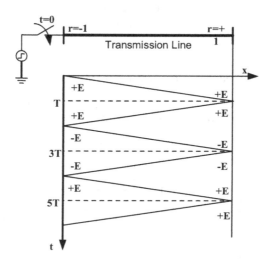

FIGURE 5.1 Lattice diagram for open circuit line energization.

DOI: 10.1201/9781003255130-5

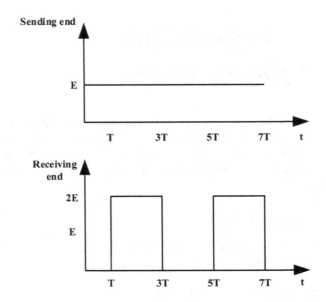

FIGURE 5.2 Sending end and receiving end voltages, open circuit line.

the line. As it is shown in the figure, the voltage at the beginning of the line is always E. The lattice diagram will provide the voltage at the end of the line.

Figure 5.2 shows that from $t = 0$ to $t = T$, which is the time taken by the wave to reach the end of the line, the voltage at the receiving end is zero. This is due to the fact that before the wave reaches the end of the line, the voltage stays zero. When $t > T$, the voltage jumps to 2E. The magnitude of the voltage stays the same until the next voltage wave reaches the receiving end for the second time (it takes 2T seconds). At $t = 3T$, which is the second arrival of the voltage wave at the receiving end, the voltage becomes zero (2E − 2E). This pattern continues, as it is suggested in Figure 5.2.

5.1.1 STEP SOURCE ENERGIZATION THROUGH INDUCTANCE

Figure 5.3 shows a transmission line that is being energized through an inductance. We can write voltage equations associated with the sending end and receiving end as follows:

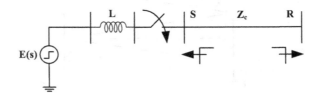

FIGURE 5.3 Line energization through an inductance.

At the energization point, that is marked with S in Figure 5.3, voltage division between the characteristic impedance and the inductor is written in Laplace domain as in Equation (5.1).

$$V_S(s) = \frac{Z_C}{Ls + Z_C} E(s) \qquad (5.1)$$

The source $E(s)$ is considered to be a step source and it can be given in Laplace domain as:

$$E(s) = \frac{E}{s} \qquad (5.2)$$

Equation (5.1) in time domain will be shown as Equation (5.3). Voltage variation at the sending end follows an exponential function. V_s is the voltage at the sending end.

$$V_S(t) = E\left(1 - e^{-\frac{Z_C}{L}t}\right) \qquad (5.3)$$

In order to use the lattice diagram in finding the voltages at points S and R, we use Heaviside theorem. In Laplace domain, by multiplying a function to e^{-Ts}, the function will be delayed by T in time domain. When the wave reaches the end of the open circuit line at $t = T$ (reflection factor is 1), we can write the voltage as:

$$V_R(s) = 2e^{-Ts}V_S(s) \qquad (5.4)$$

Lattice diagram for the abovementioned case is shown in Figure 5.4. When the wave reaches the end of the line, gets reflected, and travels toward the sending end of the line, then again, the wave gets reflected toward the end of the line and reaches the end of the line at $t = 3T$. Therefore, we can write the equation for the end of the line as:

$$V_R(s) = 2r_s e^{-3Ts}V_S(s) + 2e^{-Ts}V_S(s) \qquad (5.5)$$

Continuing the same process for the n-th encounter of the wave at the end of the line, we will have the voltage:

$$V_R(s) = (2e^{-Ts} + 2r_s e^{-3Ts} + 2r_s^2 e^{-5Ts} + \cdots)V_S(s) \qquad (5.6)$$

In Equation (5.6), r_s is found with the following:

$$r_s = \frac{Ls - Z_C}{Ls + Z_C} \qquad (5.7)$$

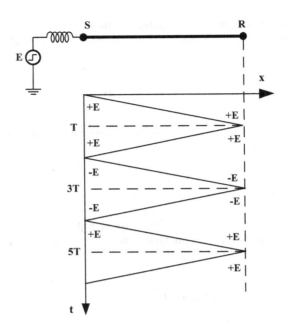

FIGURE 5.4 Lattice diagram for line energization through inductor.

If we ignore the terms for $t \geq 5T$, in Equation (5.6), we will have:

$$V_R(s) = \left[\frac{EZ_C}{(Ls + Z_C)s} \right] \left[2e^{-sT} + 2\frac{(Ls - Z_C)e^{-3sT}}{(Ls + Z_C)} \right] \tag{5.8}$$

Using the Laplace inverse, we will have the voltage at the end of the line as:

$$V_R(t) = 2u(t - T)E\left(1 - e^{\frac{Z_C(t-T)}{L}} \right) + 2u(t - 3T)E\left[-1 + e^{\frac{Z_C(t-3T)}{L}}\left(1 + \frac{2Z_C}{L}(t - 3T) \right) \right] \tag{5.9}$$

Now, we can study the receiving voltage based on the above estimation. Also, using this method, we can change the source type, as we will do in the following section by replacing the step function with a sinusoidal one.

5.1.2 Sinusoidal Source Energization through Inductance

Figure 5.5a shows a sine wave at the sending end of the line without considering any inductance after the source. In this figure, the line diagram is the source voltage, and the dotted line is the voltage at the sending end considering the effect of the inductance. The inductor is considered 0.1 H. As it can be seen in this figure, the voltage at the sending point can go beyond 1 pu (point C) due to inductance effect. Figure 5.5b shows the receiving end voltage. Here, the line diagram is the voltage at the receiving

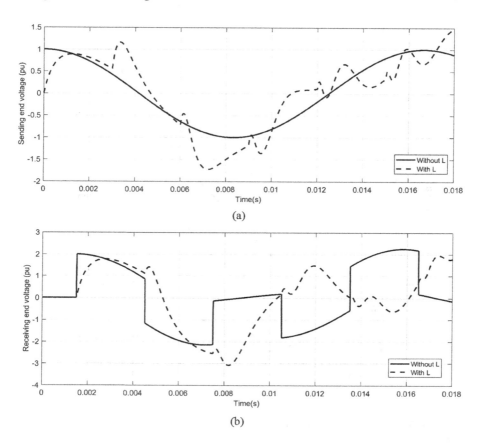

FIGURE 5.5 (a) Sending end voltage. (b) Receiving end voltage, sinusoidal source energization through an inductance.

end when the inductor is ignored, and the dotted line is the receiving end voltage while considering the 0.1 H inductor connected after the source. As it was observed in Figure 5.2, the maximum of the voltage was 2 pu. In this case, the maximum of the voltage at point D can reach to 3 pu due to inductance effect. The following diagrams are prepared in per unit scales and they are not dependent of the system rated voltage.

Figure 5.6 shows the maximum voltage at the end of the line as a function of the inductor size and also the length of the line, for the system shown in Figure 5.3. As it can be seen from the figure, as the length of the line increases, the voltage at the end of the line increases as well. At some point, this trend reverses and again goes back to the first trend, as increasing the line length increases the overvoltage at the receiving end.

Remember that an electric resonance in a circuit happens when:

$$L\omega = \frac{1}{C\omega} \tag{5.10}$$

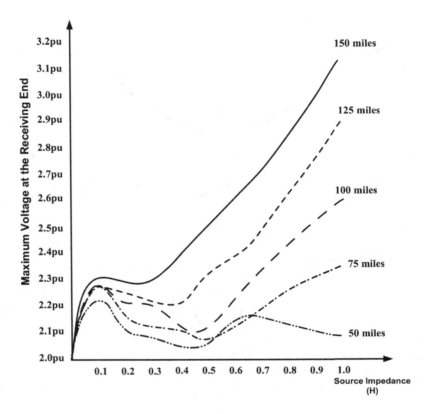

FIGURE 5.6   Maximum voltage of receiving end considering effect of source inductance and line length.

Also, the capacitive behavior of a line can reduce its inductive effect, and the amount of the capacitance in the line is dependent on the length of the line; therefore, longer line will have a larger capacitance. As it can be seen in Figure 5.6, due to higher capacitive effects of 150-mile line, reduction of the inductive behavior of this line happens in lower inductances in comparison with 100-mile line, which has a less capacitive effect.

5.1.3   LINE ENERGIZATION THROUGH ENERGIZED LINES

In this part, a line is getting energized by closing a circuit breaker to other lines that are already alive. This condition is shown in Figure 5.7. Before closing the breaker, voltage at 1 is E. At the closing time, the impedance seen at the connecting point can be written as:

$$R_e = \frac{Z_{C1} \times Z_{C2}}{Z_{C1} + Z_{C2}} \tag{5.11}$$

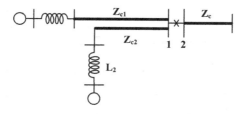

FIGURE 5.7   Line energization through energized lines.

FIGURE 5.8   Equivalent transient impedance diagram.

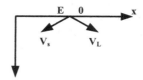

FIGURE 5.9   Forward and backward wave at the closing time.

As shown in Figure 5.8, the current going through the line will be:

$$I = \frac{E}{R_e + Z_C} \tag{5.12}$$

Analysis and studies related to transients are dependent on time and position. In this study, line characteristic impedance Z_c has a finite value as infinite impedance representsopen circuit. This would be the way of analyzing lumped circuits and not for transients. Also, the parameters R_e and Z_c in transient analysis are series components, as the current that goes through one is the same as the one that goes through the other. Figure 5.9 shows the arrangement of forward and backward waves in a lattice diagram. The following equations specify the aforementioned waves.

$$V_L = Z_C I = \frac{Z_C}{Z_C + R_e} E \tag{5.13}$$

$$V_S = -R_e I = -\frac{R_e}{Z_C + R_e} E \tag{5.14}$$

The negative sign in Equation (5.14) is due to the traveling wave direction toward the negative x-axis.

In Figure 5.7, the switch can be a circuit breaker in a high voltage switchgear. Therefore, the lines connected to the point 1 are at the same voltage level and can have the same characteristic impedance (Z_c). Now, if we assume $R_e = \dfrac{Z_c}{2}$, therefore

the magnitude for the forward wave is $V_L = \dfrac{2E}{3}$, and the magnitude for backward

traveling wave is $V_L = \dfrac{-E}{3}$. It is worth noting that the voltage at the right side of the line before closing is zero and the voltage at point 1 is E before closing. Therefore,

when the wave $V_L = \dfrac{-E}{3}$ passes through the lines with characteristic impedances of

Z_{c1} and Z_{c2}, the voltage of these lines becomes $\dfrac{2E}{3} = E - \dfrac{E}{3}$, which is the same as the voltage at the right side of the line.

5.1.4 SOURCE ENERGIZATION THROUGH INDUCTANCE ADJACENT TO PARALLEL LINES

Figure 5.10 illustrates the condition of source energization through inductance adjacent to parallel lines. This case models a local generation in a substation, and at the same time, establishes connections of transmission lines with the same substation.

Switching diagram related to Figure 5.10 is shown in Figure 5.11. In this figure, R_e is the equivalent impedance of the lines that are connected to source through L_1 and L_2 inductances.

Figure 5.12 shows the voltage at the sending end (beginning of Z_c) and receiving end of the line (end of Z_c). The dotted line shows the condition where the parallel lines are not connected, and the source is connected to an inductor energizing the line. The complete line shows the condition where the line with characteristic impedance Z_c is adjacent to two parallel lines that are connected to the source. As it can be seen, transient voltage for both receiving end and sending end is less in case of the adjacent lines.

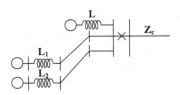

FIGURE 5.10   Source energization through inductance adjacent to parallel lines.

FIGURE 5.11   Equivalent circuit.

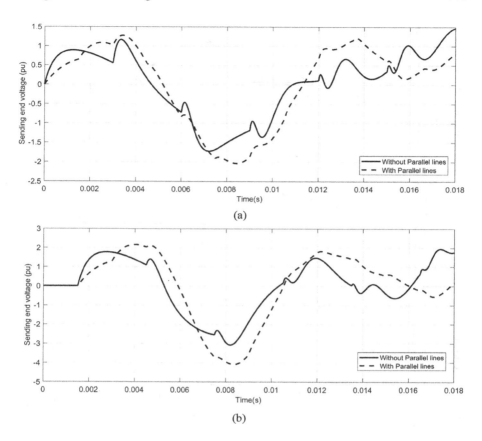

(a)

(b)

FIGURE 5.12 (a) Voltage at the sending end. (b) Voltage at the receiving end; effect of lines in parallel with source.

5.2 ASYNCHRONOUS CLOSING OF THREE PHASE CIRCUIT BREAKER CONTACTORS

In closing air pressure circuit breakers, assume that the contacts for all the three phases move and close simultaneously. Considering the 120° difference in three phase waveforms, voltage magnitudes at the closing time are different for each phase. Therefore, compared with the other two phases, one of the phases has a larger voltage difference between that phase and the other side of the contact that had zero voltage. Thus, there will be an arc between that phase and the other side of the contact. The mentioned voltage difference of other two phases is less than the first one. Therefore, the arcing phenomena makes the synchronized closing impossible. In other words, one of the phases has been energized due to the arcing and the other two are not energized yet. This situation makes an asynchronous closing with the delay of around 5 ms.

Figure 5.13 shows the receiving end voltages of phase a, b, and c. In this figure, the continuous line shows the voltage at the receiving end if the contacts are

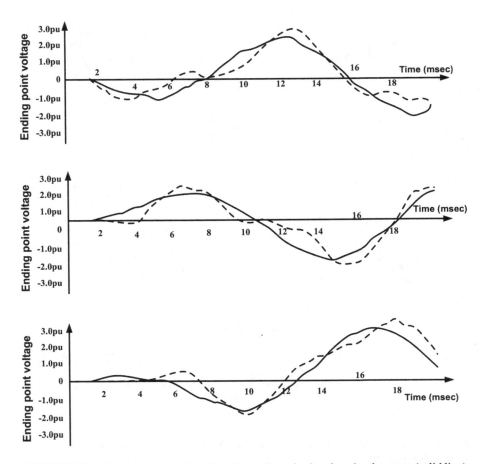

FIGURE 5.13 Receiving end voltage for phase a, b, and c, breaker simultaneous (solid line), and asynchronous (dashed line) closing.

connected simultaneously, and the dotted line shows the receiving end voltages if the contacts are closed at different times.

As it can be seen in this figure, asynchronous closing of the contacts will cause larger voltage transients at the end of the line. In order to control the overvoltages in this case, control of the closing time is implemented in the breakers. This will increase the breaker cost and it can be done in the following three ways:

    A. Closing instant control of the three phases
        The phase difference in a three-phase balanced system is 120°. Therefore, in this case, the contacts will be closed with 120° phase shift.
    B. Closing instant control with respect to the source voltage
        The breaker receives the closing command from the source at the instant where the source voltage is zero.
    C. Closing instant control considering the voltage across the breaker

Closing instant in this condition is when the voltage between the contacts of the breaker is minimal. Usually, a threshold is considered where the voltage at the breaker needs to be less than that threshold to be able to close.

5.3 REACTIVE COMPENSATION

Due to the Ferranti effect, in order to perform reactive power compensation in long transmission lines with voltages higher than 400 kV, a reactor is connected at the end of the line or beginning of the line or both. For voltages less than 400 kV, transformer tertiary winding is used to connect the reactor.

Figure 5.14 shows the maximum of the transient voltage at the end of the line in case of having a reactor at the receiving end for 50% compensation. In this figure, we have three lines with the length of 100, 200, and 240 miles. The horizontal axis presents the inductance of the source in the range of 0 to 1 H.

In Figure 5.14, for a transmission line with the length of 100 miles, and the source inductance of 1 H, 50% compensation at the end of the line will cause the maximum overvoltage at the end of the line to reach 2.1 pu. This value in Figure 5.6 is recorded as 2.6 pu. Consequently, we can see the transient overvoltage rise in both cases, but in this case, reactor installation at the end of the line has alleviated the magnitude of transient overvoltage at the end of the line.

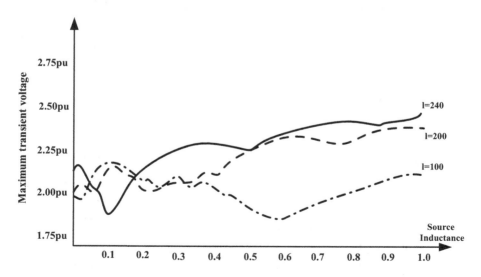

FIGURE 5.14 Receiving end maximum voltage with 50% compensation at the end of the line.

5.4 TRAPPED CHARGE

In case of a single-line to ground fault, which is the most common type of fault in power systems, the distance relays will operate at both sides of the line, and they trip all three phases of the line. In this condition, the healthy lines are also opened. For the healthy phases with no connection to the ground, the electric charges stay in those lines and these charges are called trapped charges. The line intrinsic and stray capacitors between the lines and ground will form an RC circuit, and therefore, the charges get depleted in 2–5 min. In dry weather condition, this can take up to 15 min. In case of having a reactor or voltage transformers in the line, this depletion will be done faster. In any case, the time that is required for the charge depletion is longer than the time the breakers reclose again after fault clearance. Worst case scenario is when a −1 pu trapped charge after breaker opening exists in the line, and at closing, voltage is at its highest value, which is 1 pu. This worst-case scenario is depicted in Figure 5.15. A traveling wave with the magnitude of 2 pu (the voltage across the breaker) will propagate to the line and reflection of that at the end of the line will make a wave with the magnitude of 4 pu. Now, due to the existence of −1 pu charge in the line, the voltage at the end of the line reaches 3 pu and that can make flashover condition in healthy lines. Figure 5.16 shows this condition.

Figure 5.17 illustrates the receiving end voltage of an open circuit line for two cases of trapped charge. Part a is showing the case where the system is connected to

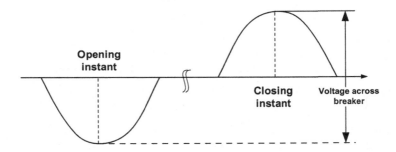

FIGURE 5.15 Worst-case scenario for reclosing a line.

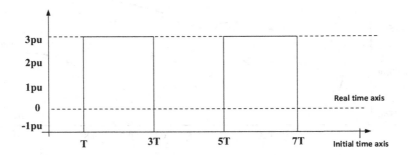

FIGURE 5.16 Voltage after reclosing, −1 pu trapped charge effect.

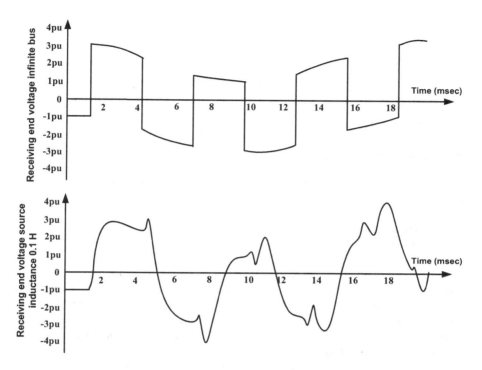

FIGURE 5.17   Receiving end overvoltage, −1 pu trapped charge case.

an infinite sinusoidal source and part b shows when the source has the fixed inductance of 0.1 H. As it is stated above, the worst-case condition happens if the trapped
charge is −1 pu, and at the closing time, we have 1 pu. As it can be seen in this figure
for case a, the transient overvoltage reaches 3 pu, and in comparison with Figure 5.5b
(continuous line), this value is higher by 1 pu. As Figure 5.17b suggests in the aforementioned worst-case scenario that if the source inductance is 0.1 H, the overvoltage
can be higher than 3 pu. The result in Figure 5.5b (dotted line) shows transient overvoltage less than 3 pu. Therefore, the system may experience a higher overvoltage
considering tapped charge for a healthy phase. In order to decrease the amount of
trapped charge in a line, an opening resistance is used as it is depicted in Figure 5.18.
This resistance is shown with parameter R. In this case, the first breaker that is labeled
1 opens, and after 30–60 ms, the breaker 2 opens. After breaker 1 opens, the line
capacitor depletes the charges through the resistance R. The time required for charge
depletion is related to the time constant of the RC circuit, which is dependent on the
length of the line (C is related to the length of the line). After opening the resistance
circuit, the line capacitor will perform depletion in a smoother way. Using voltage
transformers in the line is another way to decrease the trapped charges in the line.
These transformers provide paths to ground to discharge trapped electricity charges
to ground through the neutral grounded Y connection. Measurements show that after
0.4 s, the trapped charges are completely depleted.

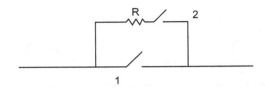

FIGURE 5.18 Opening resistance for depleting trapped charges.

5.5 SYSTEM LOSSES

So far, we have ignored the effect of losses in the system. System losses can be categorized into two parts:

- Conductor resistance and earth wire resistance
- Corona

The most important effects of the resistive losses are the attenuation of wave amplitude and also perturbation in the waveform. This happens due to the fact that frequency elements of the wave have different attenuation coefficient and different propagation speeds. Figure 5.19 shows voltage at the receiving end of a 120-mile line. The line is connected to an infinite bus, and we are considering two cases. Figure 5.19a shows voltage at the end of the line when the bottom phase is energized. Figure 5.19b shows voltage at the end of the line when the top phase is energized. Due to capacitive and inductive interactions, there will be voltage changes in other phases. This is an asymmetrical phenomenon. Therefore, we need to use the symmetrical components methodology in order to analyze this.

Figure 5.20 shows a 428 km transmission line operating at 400 kV. This line is energized through a 1 H source inductance. There is a 100 MVAr reactor at the end of the line and the line losses are considered. Figure 5.21 shows that the system losses do not always have attenuating effects. This is due to various effects on the wave properties such as different propagation speeds, phase shifts, and several reflections. These effects on the wave can cause overvoltages at the end of the line, which is depicted as a dotted line in Figure 5.21.

5.6 SWITCHING LOCATION

Consider the combination of a transmission line and a cable as shown in Figure 5.22. Assume the line that follows the 300 Ω line is an energized line. In this case, we consider two different options for switching from system operator point of view, first, energizing the combination through point A and second energizing from point B, as it is shown in Figure 5.22.

If the system is energized from point A, the traveling wave passing through point C will propagate toward the end of the overhead line, and considering the impedance shown in the above figure, we can calculate $b_c = \dfrac{2 \times 300}{30 + 300} \cong 2$, which is almost twice the wave coming from the source. The reflected wave from the end of the line will have the magnitude of almost 4 pu. Now, if the system is energized from B side, the

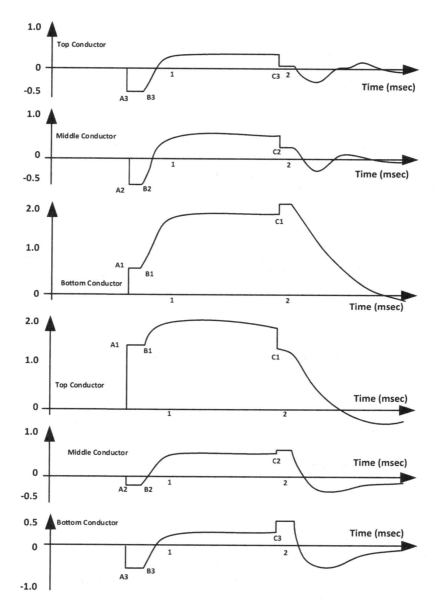

FIGURE 5.19 Applying a step function to (a) bottom phase and (b) top phase of three phase line from an infinite bus considering system losses.

wave passing through C will go through the cable with $b_c = \dfrac{2 \times 30}{30 + 300} \cong 0.2 \text{ pu}$. If the breaker A is open, the wave will be reflected with 0.4 pu. With simple calculations above, we showed that the place where system switching is being conducted has a major effect on the magnitude of transient overvoltages. In practice, the placing consideration is called "Best End Switching."

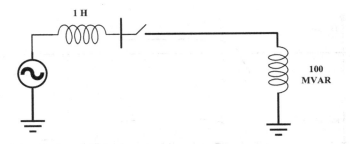

FIGURE 5.20 Transmission line energization (400 kV, 428 km) through 1 H source inductor, considering losses.

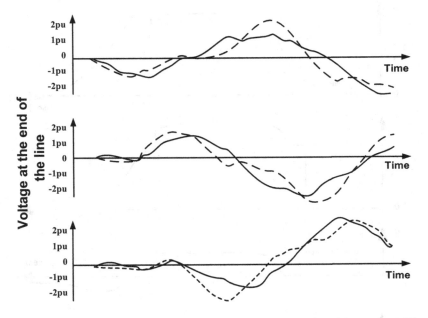

FIGURE 5.21 Receiving end voltage, no loss continuous line, and with losses dotted line.

5.7 TERMINAL TYPE OF EFFECTS

Figure 5.23 shows a 3.2-mile line that is connected to two transformers at the end of the line. In common practice, transformers are connected to load places through cables. Capacitors associated with the cables are in the range of μF in here. The transformers in their simplest form are modeled by inductances. As mentioned, in this case, we assume that there is no direct demand.

In case of having the transformers at point 3 disconnected, this case becomes exactly the same as it is seen in Figure 5.2, that is, an open circuit line. If we expand the source step function with Fourier series, we will have:

$$f(t) = E - \frac{4E}{\pi}\left[\cos\left(\frac{\pi}{2T}\right)t - \frac{1}{3}\cos\left(\frac{\pi}{2T}\right)3t + \frac{1}{5}\cos\left(\frac{\pi}{2T}\right) - \cdots\right] \qquad (5.15)$$

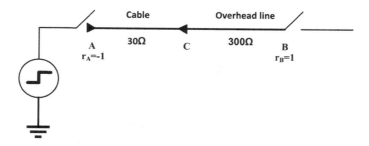

FIGURE 5.22 Switching location effect on overvoltages.

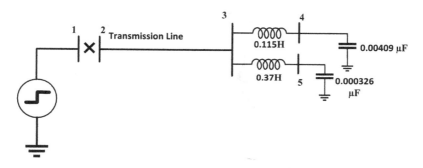

FIGURE 5.23 Transmission line connected to no load transformers.

In this case, the period of the wave is 4T; therefore, the fundamental frequency of this wave is:

$$f = \frac{1}{4T} = \frac{\upsilon}{4l} = \frac{3 \times 10^8}{4 \times 3.2 \times 1.61 \times 10^3} = 14557.5 \, \text{Hz} \tag{5.16}$$

Figure 5.24 shows the end of the line voltage. According to Equation (5.15), the input waves include different frequencies that can excite the inductor and capacitor combination at point 3. If the excitation frequency has the value $= \frac{1}{2\pi\sqrt{LC}}$, resonance phenomenon takes place. If the frequency is $\frac{1}{2\pi\sqrt{LC}} = \frac{1}{2 \times \pi \times \sqrt{0.37 \times 3.2 \times 10^{-10}}} = 14{,}491.5 \, \text{Hz}$, capacitor and inductor at point 5 are excited and that is shown in the figure. This frequency will not cause any resonance at point 4.

Figure 5.25 shows the maximum voltages of points 3, 4, and 5 as a function of the line length. Considering the relationship $f = \frac{9}{4l}$, the resonance frequency can happen at different lengths. Also at the different lengths, different harmonics of the wave as in relationship (5.15) suggests, can excite the capacitor–inductor combination. For example, for the length of 9.6 miles, the third harmonic of the exciting wave at point 3 excites this combination, which can be seen in Figure 5.25 as the second peak at l = 9.6 miles.

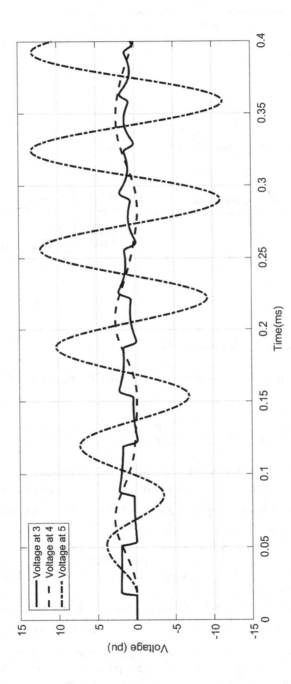

FIGURE 5.24 Voltage at points 3, 4, and 5.

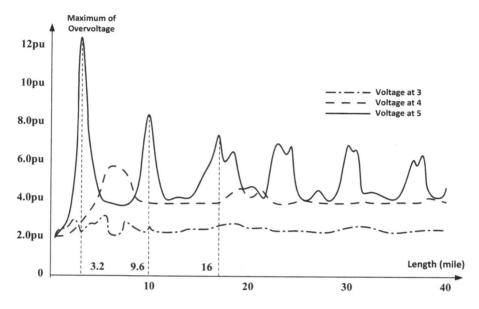

FIGURE 5.25 Voltages at points 3, 4, and 5 with different lengths of the line.

Example 5.1:   Consider the system in Figure 5.26, find voltages at A, B, C, and D until $t = 1076$ µs in the following conditions:

a.   Line CD is not supplying any loads
b.   Line CD has the trapped charge of −1 pu
c.   Line CD is not supplying any loads and there is closing resistance of 400 Ω in the breaker

For all the above cases, assume voltage at point B is 1 pu before closing the breaker. Speed of the wave is $\dfrac{1}{5.38}$ mile/µs. Ignore the changes related to 50 Hz frequency.

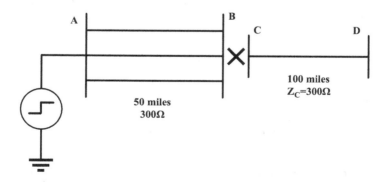

FIGURE 5.26 Line energization in different conditions.

Solution

Since we are ignoring the amplitude changes happening due to 50 Hz frequency, we need to consider only the peak of the source voltage as a step function, and we have:

$$r_{AB} = \frac{300 - 100}{300 + 100} = 0.5$$

$$b_{AB} = 1 + r_{AB} = 1.5$$

$$r_{CD} = 1$$

$$b_{CD} = 2$$

$$r_{BA} = \frac{0 - 100}{0 + 100} = -1$$

$$b_{AB} = 1 - 1 = 0$$

$$r_{DC} = \frac{100 - 300}{100 + 300} = -0.5$$

$$b_{DC} = 1 + (-0.5) = 0.5$$

We can calculate the propagation times as:

$$T_{1AB} = 5.38 \frac{\mu\sec}{\text{mile}} \cdot 50 \, \text{miles} = 269 \, \mu\sec$$

$$T_{2AB} = 5.38 \frac{\mu\sec}{\text{mile}} \cdot 100 \, \text{miles} = 538 \, \mu\sec$$

a.   Having a 1 pu voltage at B before switching, the wave propagates to both sides of the point B. Considering the fact that the line CD is open and we have three 300 Ω lines in parallel between points A and B, the forward voltage wave applied to line CD is:

$$\frac{300}{300 + 100} \times 1\text{pu} = 0.75 \, \text{pu}$$

The backward voltage wave applied to line BA is:

$$\frac{-100}{400} \times 1\text{pu} = -0.25 \, \text{pu}$$

After closing the breaker s, points B and C will become one and voltage at B is reduced to 1 − 0.25 = 0.75 pu. The results are shown in Figure 5.27. Figure 5.28 shows a graph of all the changes up to 1076 μs.

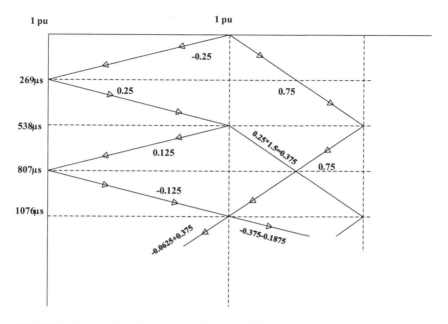

FIGURE 5.27 Lattice diagram part a Example 5.1.

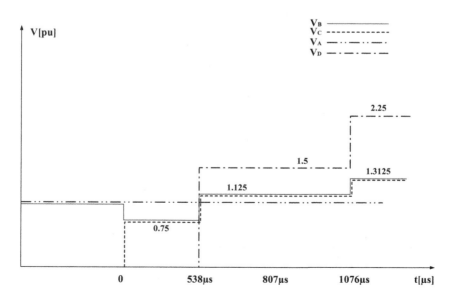

FIGURE 5.28 System all points voltage changes up to 1076 μs, part a.

b. In this case, there is −1 pu trapped charge and point B has the voltage of 1 pu. Therefore, there will be a 2 pu pulse wave across the breaker. The voltage applied to line CD will be:

$$\frac{300}{300+100} \times 2\,pu = 1.5\,pu$$

The backward voltage wave applied to line BA is:

$$\frac{-100}{300+100} \times 1\,pu = -0.5\,pu$$

After closing the breaker, the points B and C become one point and we have the lattice diagram as shown in Figure 5.29. Figure 5.30 shows the time variation of the wave at different points.

c. The series closing resistance of 400 Ω is added. Therefore, the voltage at the breaker will be:

$$\frac{400}{300+100+400} \times 1\,pu = 0.5\,pu$$

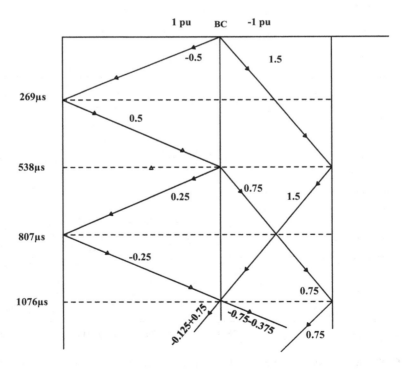

FIGURE 5.29 Lattice diagram part b Example 5.1.

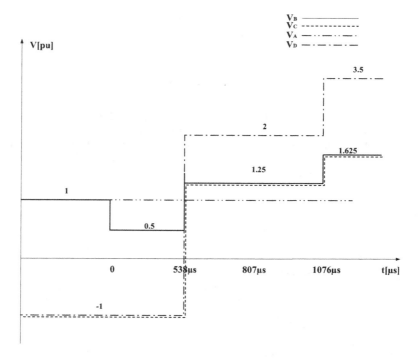

FIGURE 5.30 All points voltage changes up to 1076 μs, part b.

The voltage wave applied to line CD is:

$$\frac{300}{300 + 100 + 400} \times 1\text{pu} = 0.375\,\text{pu}$$

The voltage wave applied to line BA is:

$$\frac{-100}{300 + 100 + 400} \times 1\text{pu} = -0.125\,\text{pu}$$

The reflection and break coefficient (r and b) are:

$$r_{AB} = \frac{(400 + 300) - 100}{(400 + 300) + 100} = 0.75$$

$$b_{AB} = 1 + r_{AB} = 1.75$$

$$r_{CD} = 1$$

$$b_{CD} = 2$$

$$r_{DC} = \frac{(400 + 100) - 300}{(400 + 100) + 300} = 0.25$$

$$b_{CD} = 1 + 0.25 = 1.25$$

$$r_{BA} = -1$$

$$b_{DC} = 1 + (-1) = 0$$

Figure 5.31 is the lattice diagram and Figure 5.32 is the system voltages at points A, B, C, and D with respect to time.

It is worth noting that the wave becomes attenuated as it passes the compact closing resistance. This element is a lumped element; therefore, there is no wave propagation in this element.

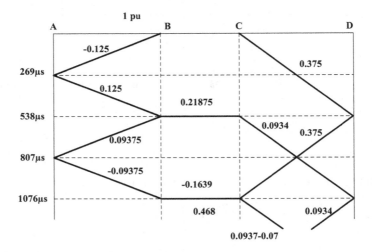

FIGURE 5.31   Lattice diagram part c Example 5.1.

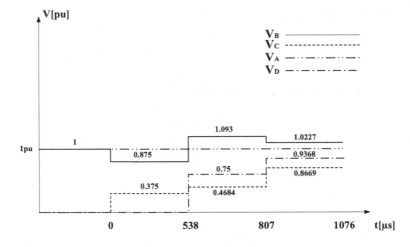

FIGURE 5.32   System voltage changes up to 1076 μs, part c.

5.8   CAPACITIVE BANK CLOSING

In order to analyze this case, we are considering Figure 5.33.
In this figure, we have the following initial conditions.

$$i_L(0) = 0 \tag{5.17a}$$

$$v_C(0) = 0 \tag{5.17b}$$

Due to the fact that in transients ωt is really small, $E\cos(\omega t)$ can be considered
E. With this approximation, the capacitor voltage is calculated as follows:

$$v_C(t) = E(1 - \cos(\omega_0 t)) \tag{5.18}$$

In this equation, $\omega_0 = \dfrac{1}{\sqrt{LC}}$.

If the system frequency f is quite smaller than the natural frequency f_0 of this circuit, voltage changes can be ignored at the closing time and a short period after closing, and the AC voltage in these studies can be considered as DC represented by E in
Equation (5.18). Figure 5.34 shows the capacitor voltage. If the voltage at connecting
time is at its maximum $(V_m = E)$, it is possible that the voltage at the capacitor bank
reaches 2 pu.

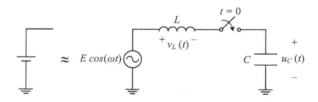

FIGURE 5.33   Capacitor bank closing circuit.

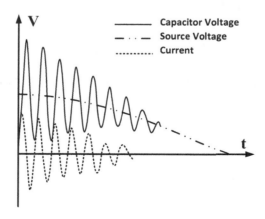

FIGURE 5.34   Capacitor voltage.

In practice, due to system losses/damping, transient voltages will damp and voltage across the capacitor becomes equal to the grid voltage. If $\omega_0 t = \pi$, then voltage at the capacitor from (5.19) is $2E$. With this, we can write the following equations:

$$v_C(t) \cong 2Eu(t) \tag{5.19}$$

$$i_C = C\frac{dv_C}{dt} = C \times 2E\frac{d}{dt}(u(t)) = C \times 2E\delta(t) \tag{5.20}$$

In this equation, $\delta(t)$ is the Dirac delta function presenting the capacitor inrush current at the closing time. However, taking the derivative of Equation (5.18), we have:

$$i_C = C\frac{dv_C}{dt} = CE\omega_0 \sin(\omega_0 t) \tag{5.21}$$

According to this equation, the capacitor maximum current is: $E\sqrt{\dfrac{C}{L}}$

With Laplace transformation and writing the KVL for the circuit, we will have:

$$\vec{I} = \frac{\vec{E}}{jL\omega + \dfrac{1}{jC\omega}} = \frac{jE}{\dfrac{1}{c\omega} - L\omega} = \frac{jEc\omega}{1 - Lc\omega^2} \tag{5.22}$$

Knowing the fact that $\vec{E}$ is the network voltage and $E = E\cos(\omega t)$, so we can write:

$$\vec{I} = \frac{Ec\omega \cos(\omega t + 90)}{1 - Lc\omega^2} = \frac{Ec\omega \sin(\omega t)}{1 - Lc\omega^2} \tag{5.23}$$

Capacitor voltage can be written as:

$$\vec{V_C} = \frac{1}{jC\omega}\vec{I} = \frac{E}{1 - Lc\omega^2} \tag{5.24}$$

The above equation shows the capacitor voltage in the Laplace domain. Using Laplace inverse, we will have:

$$v_C(t) = \frac{E\cos(\omega t)}{1 - Lc\omega^2} \tag{5.25}$$

Initially, the current can be written as:

$$i(t) \cong \frac{EC\omega t}{1 - Lc\omega^2} + \sqrt{\frac{C}{L}}E\sin(\omega_0 t) \tag{5.26}$$

If $t = \dfrac{\pi}{2\omega_0}$, then the maximum value of the current will be:

$$\max i(t) \simeq \frac{EC\omega \dfrac{\pi}{2\omega_0}}{1 - Lc\omega^2} + \sqrt{\frac{C}{L}}E \tag{5.27}$$

At the end, the voltage at the inductor can be calculated as follows:

$$v_L(t) = L\frac{di}{dt} = E_m \cos(\omega t) - u_C(t) \tag{5.28}$$

5.9 PSCAD EXAMPLES

5.9.1 SINUSOIDAL SOURCE ENERGIZATION THROUGH INDUCTANCE

We continue in this part by providing simulation cases related to the energization cases discussed in this chapter. The case study related to the subject is illustrated in Figure 5.35. In this simulation, the source is a 150 kV at 60 Hz. The surge impedance

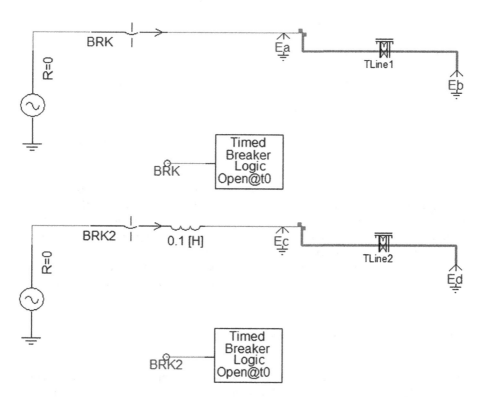

FIGURE 5.35 Simulation system for sinusoidal source energization (cosine input) through inductance. (top a, bottom b).

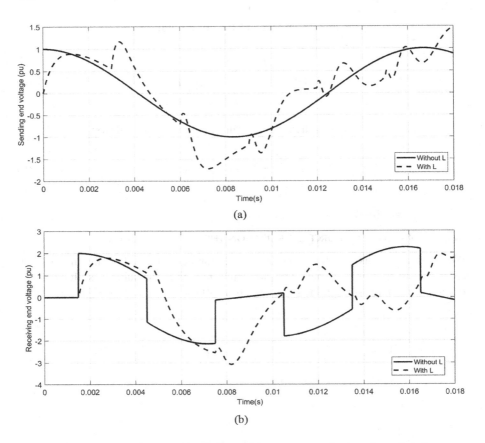

FIGURE 5.36 Simulation results of sinusoidal source energization through inductance: (a) sending end voltages and (b) receiving end voltages ($V = \cos(\omega t)$).

of the line is 240 Ω with a traveling time of 1.5 ms. Circuit a is energized without inductors and circuit b is energized through a 0.1 H inductor. Figure 5.36a shows the sending end voltages of the conditions with and without inductor and Figure 5.36b shows the respective end of the line voltages. In case of having the inductor, the mentioned figures show the overvoltages on the voltage wave.

Figures 5.37a and 5.37b are repeating the abovementioned simulations. The only difference is that the new simulations are done with sine wave input (5.36a and b are simulated using a cosine input).

5.9.2 SOURCE ENERGIZATION THROUGH INDUCTANCE ADJACENT TO PARALLEL LINES

In this section, a simulation of a line energization adjacent to parallel lines is studied. The simulation circuit is illustrated in Figure 5.38. Source is a 150 kV at the frequency of 60 Hz. System contains a transmission line with the impedance of 240 Ω

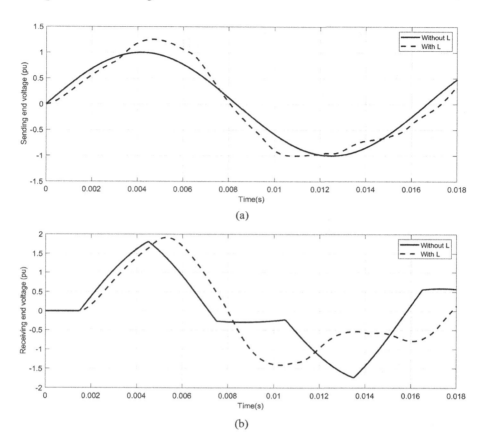

FIGURE 5.37 Simulation results of sinusoidal source energization through inductance:
(a) sending end voltages and (b) receiving end voltages ($V = \sin(\omega t)$).

and a traveling time of 1.5 ms. Two different conditions are considered; the first
condition is energization through a 0.1 H without the parallel lines and the second
condition is energization with the parallel lines. In the second condition, characteristics of the parallel line are similar to the main line characteristic. The parallel lines
are energized through inductances of 0.35 and 0.4 H.

Figure 5.39a shows voltage at the sending end for conditions with and without the
parallel lines. Similarly, Figure 5.39b illustrates voltages at sending end with and
without parallel lines conditions. In case of having the mentioned 0.1 inductor, overvoltages will occur at the sending end and receiving end of the line. Also, the results
show overvoltages in case of having the mentioned parallel lines.

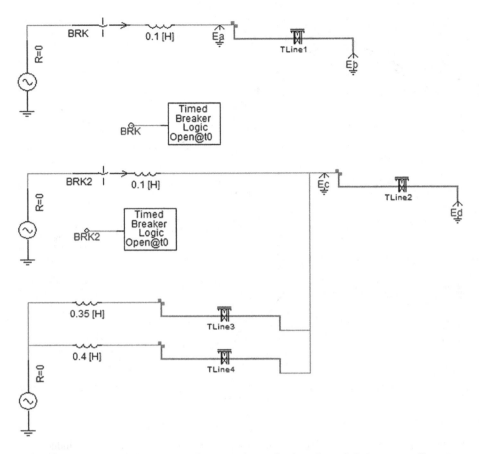

FIGURE 5.38 Simulation system for source energization through inductance adjacent to parallel lines.

5.9.3 Terminal-type Effects

In this part, we analyze the condition of line energization with two transformers and cables at the end of the line. Circuit diagram of the simulation is presented in Figure 5.40. Source is 150 kV DC input. Surge impedance of the line is 240 Ω with a traveling time of 17 μs. The transformers and the cables at the end of the line are modeled with inductors and capacitors as illustrated in Figure 5.40. In Figure 5.41, end of the line voltage (E3) and voltage at the capacitor–inductor connection point voltages (E4 and E5) are illustrated. As it can be seen that due to the resonance phenomena, there is a possibility of intensive overvoltages at (E5).

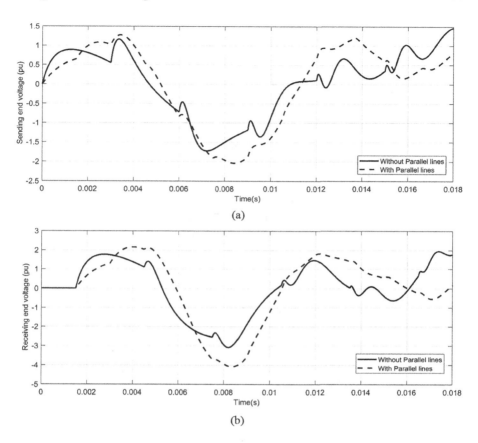

FIGURE 5.39 Simulation results of source energization through inductance adjacent to parallel lines: (a) Sending end voltages and (b) receiving end voltages.

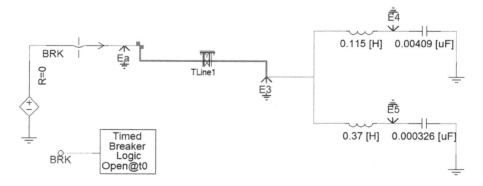

FIGURE 5.40 Simulation system for considering two transformers at the receiving end.

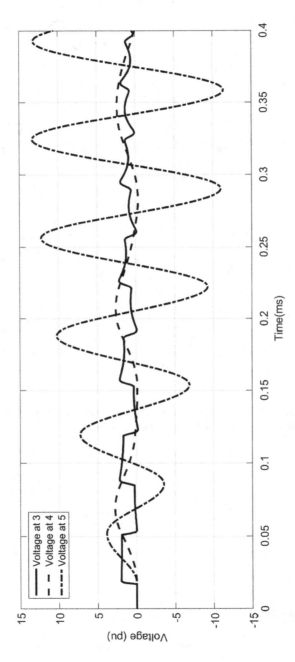

FIGURE 5.41 Simulation results of considering two transformers at the receiving end.

Problems

5.1 After a single phase to ground fault happened in a three-phase system, the breaker is tripped at the faulty line. When the breaker opens, voltage is at its maximum and there is trapped charge in the line that decays with the time constant of 60 ms (i.e., $v_B = V_0 e^{-377t} msec$). The circuit breaker is equipped with reclosing mechanism and it recloses after 30 ms. Find the maximum voltage appearing at the end of the line, ignoring wave reflections at the beginning of the line.

5.2 In the following circuit at $t = 0$, breaker K closes.
   a. Assuming that there is no load at the end of the line. Find the voltage waveform at the end of the line (t).
   b. If the LC load is connected at the end of the line. Assess the possibility of having overvoltages for point w.
      Assume $C = 40\ nF$, $L = 1mH$, and $Z_A = Z_B$.

5.3 For the following system, find the best time to close the circuit breakers synchronously to minimize the voltage pressure between the minimal phases.

5.4 Consider the following figure.
   a. If there are no inductors and no capacitors at point 3, plot $v_3(t)$ in case of closing the breaker between buses 1 and 2.
   b. In case of having all the inductors and capacitors connected at point 3 and closing the breaker, waveforms for voltages v_3, v_4, and v_5 can be seen in the following figure.
      Explain why the v_5 is showing larger changes in the magnitude.
   c. If length of the line (1) changes, how would that affect the v_3, v_4, and v_5?

5.5 Consider a system containing a 265-mile line connected to a source with a constant inductance. Record magnitudes of overvoltages at the receiving end in case of installation of a 100 MVAr reactor and/or 200 MVAr reactor at sending end and/or receiving end of the line. Fill out the following table (Table p5.1). Is it possible to draw a logical relationship between the installation location and the magnitude of reactors to the magnitude of the observed overvoltages? Use PSCAD file supplied for problem 5 of Chapter 5 to conduct the experiment.

Check the following statements and make proper comments for each of them:

   a. Overvoltage relationship to increase of source inductance
   b. Magnitude of installed reactor (100 or 200 MVAR) and their effect on transient overvoltages
   c. Location of the reactor and its effect on transient overvoltages
   d. Increasing the size of the reactor at the sending end, decrease the magnitude of transient overvoltage

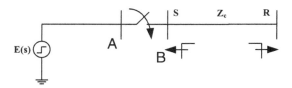

FIGURE P5.1   Line energization, trap charge effect.

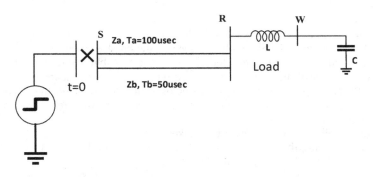

FIGURE P5.2   Line energization, load effect.

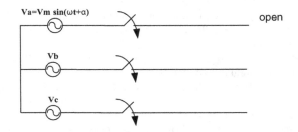

FIGURE P5.3   Circuit breakers synchronous line energization.

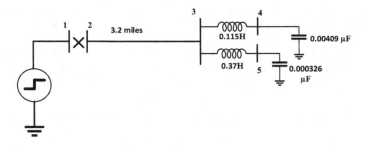

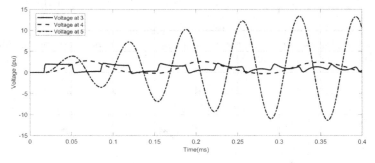

FIGURE P5.4   Possible resonances in different nodes.

TABLE P5.1
Overvoltages at Receiving End, Effect of Reactor Installation at Sending and/or Receiving Ends

Source Inductance (H)	Voltage at the End of the Line NO Reactor	Voltage at the End of the Line (pu) with 100 MVAR Reactor Installation at:			Voltage at the End of the Line (pu) with 200 MVAR Reactor Installation at:	
		Sending End	Receiving End	Both	Sending End	Receiving End
0						
0.05						
0.1						
0.5						

BIBLIOGRAPHY

1. Neville Watson and Jos Arrillaga, *Power Systems Electromagnetic Transients Simulation*, London, United Kingdom: The Institution of Engineering and Technology, 2007.
2. C. L. Wadhwa, *Electrical Power Systems*, New Academic Science, 2011.
3. Lou van der Sluis, *Transients in Power Systems*, John Wiley & Sons, 2001.
4. Juan A. Martinez-Velasco, *Transient Analysis of Power Systems: A Practical Approach*, John Wiley & Sons, 2019.
5. Juan A. Martinez-Velasco, *Power System Transients: Parameter Determination*, The United States of America: Taylor & Francis Group, 2010.
6. Yoshihide Hase, *Handbook of Power System Engineering*, John Wiley & Sons, 2007.
7. J. C. Das, *Transients in Electrical Systems: Analysis, Recognition, and Mitigation*, The United States of America: Ma Graw Hill, 2010.
8. N. Mohan, W. P. Robbins, T. M. Undeland, R. Nilssen, and O. Mo, "Simulation of Power Electronic and Motion Control Systems – An Overview", *Proceedings of the IEEE*, Vol. 82, No. (8), 1994, pp. 1287–1302.
9. S. H. Hosseini, G. B. Gharehpetian, and B. Vahidi, "IPC Transformer Winding VFTO Control by Using Closing Resistance", *19-th International Power System Conference*, 22–24 November 2004, Tehran, Iran (in Persian).
10. Mohammad Jazini, S. M. Mousavi, G. B. Gharehpetian, and Mohssen Rafie, "Increasing of Contact Strength with Nano-materials in Electrical Breakdown", *2nd Regional CIRED Conference and Exhibition on Electricity Distribution*, January 14–15 2014, Tehran, Iran.
11. E. Goodarzi, G. B. Gharehpetian, M. Ghorat, and B. Moradzadeh "Effects of ZnO Application in Circuit Breakers for TRV Reduction", *21-th International Power System Conference*, November 13–15, 2006, Tehran, Iran (in Persian).
12. F. Shahnia and G. B. Gharehpetian, "Lightning and Switching Transient Overvoltages in Power Distribution Systems Feeding DC Electrified Railways", *3-rd International Conference on Technical and Physical Problems in Power Engineering (TPE-2006)*, 29–31 May 2006, Ankara, Turkey.
13. J. P. Bickford, N. Mullineux, and J. R. Reed, "Computation of Power System Transients", *IEE Monograph Series*, Vol. 18, 1976.176 pages.

6 Transients Induced by De-energization

In this chapter, we analyze the transients that resulted from breaking energized lines and system apparatus. Also, we will model the de-energization transients using experimental and operational assumptions. De-energization can happen due to fault clearance or intentional load or device electricity interruption.

6.1 TRANSIENT RECOVERY VOLTAGES

Consider Figure 6.1, which is suitable in analyzing the concept of transient recovery/restriking voltage, which in short is called TRV. The capacitor shown in the figure is the capacitive charge between the breaker and the earth or between the busbar and the earth, and it is not part of the equipment installed in the system. But this capacitance has a considerable impact on the analysis of transients in the system. In this figure, before opening the breaker, voltage across the capacitor is almost zero due to the fault, and after opening the breaker, the capacitor voltage becomes the same as the voltage at the breaker. KVL in the circuit provides:

$$V_m \cos(\omega t) = L\frac{di}{dt} + V_C \tag{6.1}$$

The relationship between current and voltage in the capacitor is:

$$i = C\frac{dV_C}{dt} \tag{6.2}$$

Eliminating current in Equations (6.1) and (6.2), we will have a second-order differential equation as:

$$\frac{dV_C^2}{dt^2} + \frac{V_C}{LC} = \frac{V_m}{LC}\cos(\omega t) \tag{6.3}$$

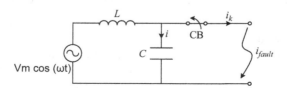

FIGURE 6.1 Fault opening equivalent circuit.

DOI: 10.1201/9781003255130-6

Defining $\omega_0^2 \triangleq \dfrac{1}{LC}$ and using Laplace transformation, we will have:

$$V_C(s) = \omega_0^2 V_m \frac{s}{(s^2+\omega^2)\times(s^2+\omega_0^2)} + V_C(0)\frac{s}{s^2+\omega_0^2} + \frac{V_C'(0)}{s^2+\omega_0^2} \tag{6.4}$$

Considering zero voltage and current at $t = 0$ for the capacitor, we can write the initial conditions as:

$$V_C(0) \simeq 0$$

$$V_C'(0) = \frac{1}{C}i(0) = 0 \tag{6.5}$$

Applying the initial condition and regrouping, we will have:

$$V_C(s) = V_m \frac{\omega_0^2}{\omega_0^2-\omega^2}\times\left[\frac{s}{s^2+\omega^2} - \frac{s}{s^2+\omega_0^2}\right] \tag{6.6}$$

With an inverse Laplace transform, we can write:

$$V_C(t) = V_m \frac{\omega_0^2}{\omega_0^2-\omega^2}\times[\cos(\omega t)-\cos(\omega_0 t)] \tag{6.7}$$

Knowing $\omega_0 \gg \omega$, Equation (6.7) changes to:

$$V_C(t) = V_m[1-\cos(\omega_0 t)] \tag{6.8}$$

As we mentioned, the voltage across the breaker is the same as the voltage across the capacitor after opening. Equation 6.8 provides the voltage across the breaker after opening as well.

Figure (6.2) illustrates the voltage across the breaker. At instant $t = t_0$, the breaker receives the opening command from protection system and the breaker contact starts its movement (opening). The current cannot be interrupted immediately considering the fault current flow. Therefore, there will an arc between contacts of the breaker. This arc has a resistive behavior and results in a small voltage drop in the period of $[t_0, t_1]$, as shown in Figure 6.2. At that instant, where the ac current becomes zero at $t = t_1$, breaking mechanism of the breaker will turn the arc off. At this instant, the voltage across the breaker follows Equation (6.8). The breaker voltage will oscillate between zero and twice the maximum of the voltage with the frequency of $\dfrac{1}{2\pi\sqrt{LC}}$.

Figure 6.3 shows the TRV phenomenon happens on the breaker voltage. This figure has been obtained by scaling Figure 6.2.

It is well-known that the magnitude of voltage in an airgap between two electrodes and the rate of rise of voltage are two important factors in insulation breakdown occurrence. This fact is presented by a curve in "breakdown voltage (y-axis)-electrodes airgap distance (x-axis)." In case of having a moving electrode with the speed of v, we can divide the x-axis, that is, airgap distance, to $v.$, and as a result, this axis will represent the time as shown in Figure 6.4.

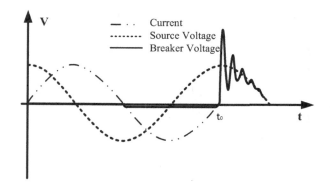

FIGURE 6.2   Breaker voltage.

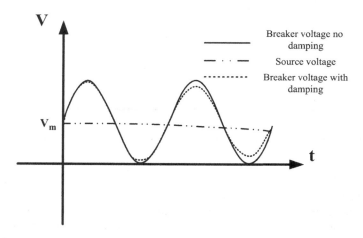

FIGURE 6.3   Breaker voltage at the instant of TRV.

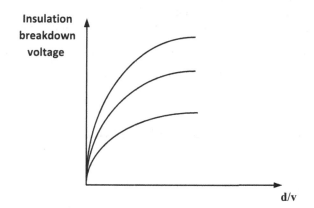

FIGURE 6.4   Airgap insulation withstand versus time ($t = d/v$)

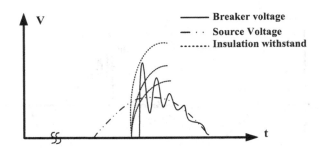

FIGURE 6.5 Insulation withstand curve and breaker voltage.

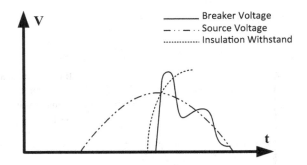

FIGURE 6.6 Lower natural frequency and less arcs.

In Figure 6.5, we are overlapping Figure 6.4 with the voltage across the breaker (Figure 6.2) in time domain, that is, using the same time axis. As it can be seen in this figure, in places where the breaker voltage is higher than the withstand voltage of the airgap, there will be an arc "Reignition." This phenomenon can happen several times when opening the circuit breaker and eventually may damage the breaker. Therefore, it is required to have the withstand curve [dotted lines in Figure 6.5] above the breaker voltage in order to prevent the TRV phenomena. This issue must be considered in design and utilization phases of a circuit breaker.

Natural frequency of the circuit plays an important role in Reignition phenomena. In TRV, the frequency of the voltage across breaker is $\dfrac{1}{2\pi\sqrt{LC}}$; therefore, higher frequencies (lower Cs) can cause several arcs and are opposite for lower frequencies, as it is shown in Figure 6.6.

6.2 RATE OF RISE OF RECOVERY VOLTAGE

Rate of rise of recovery voltage or RRRV is defined as:

$$RRRV = \frac{MaxTRV}{\dfrac{T_0}{2}} \tag{6.9}$$

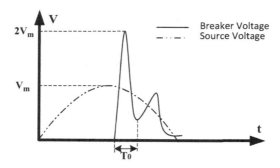

FIGURE 6.7 Period of TRV.

In Equation (6.9), T_0 is the period of the TRV as shown in Figure 6.7. Considering the fact that the maximum of TRV is twice the maximum of the source voltage, we can rewrite Equation (6.9) as follows:

$$RRRV = \frac{MaxTRV}{\frac{T_0}{2}} \quad RRRV = \frac{MaxTRV}{\pi\sqrt{LC}} \tag{6.10}$$

Also, we know $T_0 = \dfrac{1}{f_0} = 2\pi\sqrt{LC}$.

6.3 CURRENT INJECTION METHODOLOGY

In circuit theory, we have learned that a faulty branch can be replaced with a current source. The current of this source must be equal to the fault current of that branch. In order to model the opening of this branch, it is enough to inject the same current source in an opposite direction. Based on this approach, known as the current injection method, we can solve the equations of the opening of a fault. The current injection method is the superposition of circuit response in two different conditions before breaker disconnection and after disconnection. As shown in Figure 6.8, the disconnection should happen because of a fault in the circuit. Before breaker disconnection due to the fault current, a current equal to the fault current is modeled/injected in/to the faulty branch (see Figure 6.8), and after disconnection, the same current with the opposite direction is considered to be injected into the branch nodes. The final result is the superposition of circuit response in two cases.

In case of breaker opening shown in Figure 6.1, when the fault occurs, the capacitor voltage and the current that passes through it are zero. The fault current through the circuit breaker in Laplace domain is:

$$I_{CB}(s) = \frac{V_m}{L\omega} \times \frac{\omega}{s^2 + \omega^2} \tag{6.11}$$

Therefore, the current being injected in an opposite direction is determined. Now, consider Figure 6.9 to calculate the voltage at the breaker when the fault is interrupted, using this current injection.

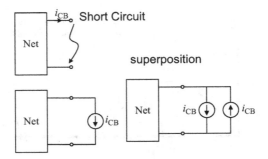

FIGURE 6.8   Analysis of system disconnection with current injection method.

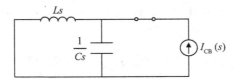

FIGURE 6.9   Equivalent circuit in current injection method, injection in an opposite direction.

Considering the parallel combination of the inductor and the capacitor, the voltage at the breaker can be calculated as:

$$V_{CB}(s) = V_m \frac{1}{s^2 + \omega^2} \times \frac{s}{1 + LCs^2} \tag{6.12}$$

We apply Laplace inverse on Equation (6.12) with the following steps:

$$V_{CB}(s) = V_m \times \left[\frac{s}{s^2 + \omega^2} - \frac{s}{s^2 + \omega_0^2} \right] \times \frac{\omega_0^2}{\omega_0^2 - \omega^2} \tag{6.13}$$

In this equation, $\omega_0^2 = \dfrac{1}{LC}$. The Laplace inverse will give us the voltage at the breaker in time domain as given in Equation (6.14).

$$V_{CB}(t) = V_m \times \frac{\omega_0^2}{\omega_0^2 - \omega^2} \times [\cos(\omega t) - \cos(\omega_0 t)] \tag{6.14}$$

The voltage at the breaker is the superposition of the voltages before breaking, which was zero, and after breaking. Therefore, the final voltage on the breaker is calculated by Equation (6.14). It is obvious that Equations (6.14) and (6.7) will provide the same result for the voltage across the breaker after breaking.

6.4 FACTORS AFFECTING TRANSIENT RECOVERY VOLTAGES

6.4.1 POWER FACTOR EFFECT ON TRANSIENT RECOVERY VOLTAGES

Figure 6.10 shows TRV for different power factors. As it can be seen that as the power factor approaches 1, the TRV will decrease. Equation (6.7) can help in analyzing this situation. In this equation as the power factor approaches 1, in disconnection time, the voltage is less than the maximum of the system voltage.

6.4.2 NATURAL FREQUENCY EFFECT ON TRANSIENT RECOVERY VOLTAGES

Natural frequency of the circuit is $f_0 = \dfrac{1}{(2\pi\sqrt{LC})}$, and according to Equation (6.8), the value of $\cos(\omega t)$ is close to 1 as the natural frequency increases. Therefore, the maximum of overvoltage can reach as high as twice the maximum of the voltage. Also, as the natural frequency grows, the period of the wave decreases. Equation (6.9) suggests that this condition will cause an increase in TRV. Figure 6.11 shows this condition.

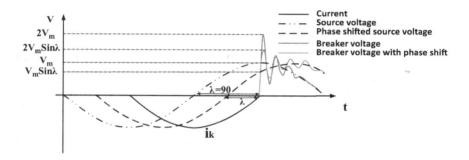

FIGURE 6.10 Power factor effect on TRV.

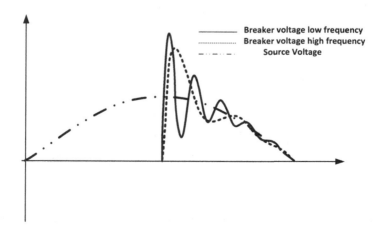

FIGURE 6.11 Natural frequency effect on transient recovery voltages.

6.4.3 EFFECT OF EXISTENCE OF TWO NATURAL FREQUENCIES ON TRANSIENT RECOVERY VOLTAGES

Figure 6.12 shows a single-line diagram of a system containing equivalent circuit of combination of a source, transmission line, and a no-load transformer. Here, we are assessing the de-energization of no-load transformer. We are using Laplace transform and current injection methodology in order to assess that.

Before opening the breaker, the current flowing through it is:

$$i_{CB}(t) = \frac{V_m \sin(\omega t)}{(L_1 + L_2)\omega} \qquad (6.15)$$

In order to use the current injection methodology, we present the current through the breaker using Laplace transform:

$$I_{CB}(s) = \frac{V_m}{(s^2 + \omega^2)(L_1 + L_2)} \qquad (6.16)$$

The voltage at the breaker is:

$$V_{CB}(s) = Z_P(s) \times I_{CB}(s) \qquad (6.17)$$

In the above equation, Z_p can be written as:

$$Z_P(s) = \left(L_1 s \,\|\, \frac{1}{C_1 s} \right) + \left(L_2 s \,\|\, \frac{1}{C_2 s} \right) \qquad (6.18)$$

Considering the fact that in transient analysis, the time to investigate the system response is really short; therefore, we can use the approximation of $\sin(\omega t) \approx \omega t$.

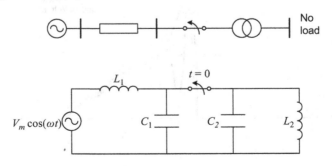

FIGURE 6.12 Single-line diagram of a no-load transformer de-energization (top) and equivalent circuit (bottom).

Current passing through the breaker before de-energization can be written in the Laplace domain and time domain as follows:

$$i_{CB}(t) = \frac{V_m}{L_1 + L_2} t \tag{6.19}$$

$$I_{CB}(s) = \frac{V_m}{L_1 + L_2} \times \frac{1}{s^2} \tag{6.20}$$

The voltage across the breaker is:

$$V_{CB}(s) = \frac{V_m}{s^2 \times (L_1 + L_2)} \left[\frac{L_1 s}{L_1 C_1 \times \left(s^2 + \dfrac{1}{L_1 C_1} \right)} + \frac{L_2 s}{L_2 C_2 \times \left(s^2 + \dfrac{1}{L_2 C_2} \right)} \right] \tag{6.21}$$

Defining $\omega_1 \triangleq \dfrac{1}{L_1 C_1}$ and $\omega_2 \triangleq \dfrac{1}{L_2 C_2}$ and implementation of Laplace transform in the time domain, we will have:

$$V_{CB}(t) = \frac{V_m}{(L_1 + L_2)} [L_1 \times (1 - \cos(\omega_1 t)) + L_2 \times (1 - \cos(\omega_2 t))] \tag{6.22}$$

As it can be seen with the abovementioned approximation, the 50 Hz alterations have not been considered in Equation (6.22).

Let us consider the following two conditions:

First condition if $\omega_1 \gg \omega_2$:

The voltage waveform across the breaker is as shown in Figure 6.13. In this figure, two different oscillations, the 50 Hz alterations and damping effects, are obvious.

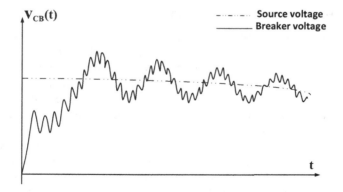

FIGURE 6.13 Two natural frequencies effect on TRV, $\omega_1 \gg \omega_2$.

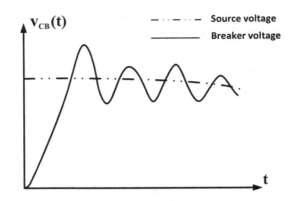

FIGURE 6.14   Two natural frequencies effect on TRV, $\omega_1 \simeq \omega_2$.

Second condition if $\omega_1 \simeq \omega_2$:

Figure 6.14 shows the voltage at the breaker before de-energization. As it can be seen, the breaker voltage in this condition is similar to the condition of having only one natural frequency, but with more distortion compared with the case with one natural frequency.

6.4.4   DAMPING EFFECT ON TRANSIENT RECOVERY VOLTAGES

So far, we have ignored the damping effect in our studies. If we consider the skin and proximity effects, using the knowledge gained in Chapter 2, we can conclude that with the increase in system losses, the magnitude of transient voltages will decrease. Therefore, no loss condition will provide us the worst-case scenario in analyzing the system overvoltages.

6.4.5   EFFECT OF TYPE OF SHORT CIRCUIT

Figure 6.15a shows a three phase to ground fault, which will be studied in this subsection. Due to the fact that three-phase voltages have different phase values at any instant, one of the contacts will operate faster than the other two. In this condition, two of the phases are closed and connected to the fault as shown in Figure 6.15a, and one phase is opened. As shown in this figure, point X in the opened circuit has the voltage of phase a, and on the other side of the opened breaker, the voltage is equal to the voltage of midpoint between two phases b and c. Point X in Figure 6.15b shows the maximum of voltage on phase a, and point Y is the midpoint between two phases b and c. Therefore, the voltage across breaker (voltage difference between X and Y) is 1.5 times the phase voltage. Since the maximum of TRV is twice the voltage maximum, therefore, the maximum of TRV can be as high as three times the maximum phase voltage.

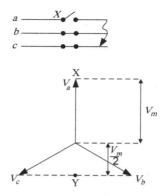

FIGURE 6.15 (a) Three phase to ground fault and (b) phase displacement and unsymmetrical breaker operation.

6.4.6 Arc Voltage Effect on TRV

As shown in Figure 6.2, before the breaker opens at instant $t = t_1$, due to fault current flow through breaker contacts and resulting arc, we have a small voltage, which is not zero. To consider the effect of this arc, the breaker voltage can be rewritten using Equation (6.4), as follows:

$$V_{CB}(t) = V_m[1 - \cos(\omega_0 t)] + V_{CB}(0)\cos(\omega_0 t) \tag{6.23}$$

We can rewrite the above equation as:

$$V_{CB}(t) = V_m - [V_m - V_{CB}(0)]\cos(\omega_0 t) \tag{6.24}$$

According to the above equation, the maximum of TRV happens when $\omega_0 t = \pi$. So, we can write the maximum of TRV as:

$$|V_{CB}(t)| = 2V_m - V_{CB}(0) \tag{6.25}$$

The $V_{CB}(0)$ is a negative number, as shown in Figure 6.2. Therefore, maximum value of TRV is larger than twice the maximum value of phase voltage.

6.5 FAULT INTERRUPTION IN A SHORT LINE

Figure 6.16 shows a line with a fault. Assume that the fault happens at x distance from the beginning of the line, which has the inductance of L_L per unit length.

In this case, the short circuit current can be written as:

$$i_{CB} = \frac{E_m \sin(\omega t)}{\omega(L_S + xL_L)} \tag{6.26}$$

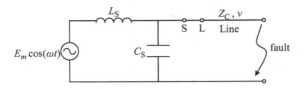

FIGURE 6.16 Fault clearing in a short line.

In this equation, we have neglected the effect of C_s before the breaker opening. The value of C_s can be in the range of nF. Therefore, in power frequency, the reactance shown by the capacitor branch is really large so this branch can be considered an open circuit. In fast transient analysis, since t is small, for example, in the range of microsecond, we can use the approximation of $\sin(\omega t) \approx \omega t$. Therefore, we will have:

$$i_{CB} \approx \frac{E_m t}{L_S + xL_L} \tag{6.27}$$

Using Laplace transform, we have:

$$I_{CB}(s) = \frac{E_m}{L_S + xL_L} \times \frac{1}{s^2} \tag{6.28}$$

Using the current injection methodology, voltage at point S, in Figure 6.16, can be written as:

$$V_S(s) = \frac{L_S s}{1 + L_S C_S s^2} I_{CB}(s) \tag{6.29}$$

This voltage in time domain will be:

$$V_S(t) = \frac{E_m L_S}{L_s + xL_L} [1 - \cos(\omega_1 t)] \tag{6.30}$$

In the last equation, $\omega_1 \triangleq 1 / \sqrt{L_s C_s}$.

Before opening the breaker, the voltage at S is:

$$V_S(t) = \frac{E_m x L_L}{L_s + xL_L} \tag{6.31}$$

Therefore, the complete response of the circuit at point S will be obtained by:

$$V_S(t) = \frac{E_m x L_L}{L_s + xL_L} + \frac{E_m L_S}{L_s + xL_L} [1 - \cos(\omega_1 t)] \tag{6.32}$$

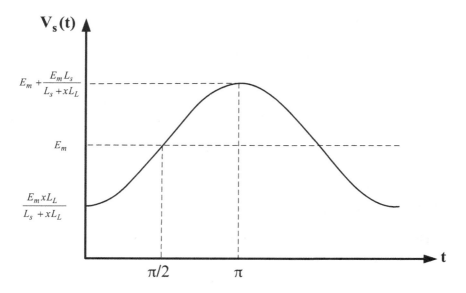

FIGURE 6.17   Voltage at point S.

The voltage at S is depicted in Figure 6.17.
Voltage at point L before opening the breaker is:

$$V_L(t) = V_S(t) = \frac{E_m x L_L}{L_s + x L_L} \tag{6.33}$$

After opening the breaker voltage at point L can be found as:

$$V_L(t) = -i_{CB}(t) \times Z_C = \frac{-E_m t}{L_S + x L_L} \times Z_C \tag{6.34}$$

The wave after opening the breaker travels to the end of the line, and later, it will be reflected back from the short circuit to the point L as:

$$V_L(t) = \frac{-E_m Z_C}{L_S + x L_L} \times \left[t \times u(t) - 2\left(t - \frac{2x}{v}\right) \times u\left(t - \frac{2x}{v}\right) + 2\left(t - \frac{4x}{v}\right) \times u\left(t - \frac{4x}{v}\right) + \cdots \right] \tag{6.35}$$

Complete voltage response at point L will be:

$$V_L(t) = \frac{-E_m Z_C}{L_S + x L_L} \times \left[t \times u(t) - 2\left(t - \frac{2x}{v}\right) \times u\left(t - \frac{2x}{v}\right) + \cdots \right] + \frac{E_m x L_L}{L_S + x L_L} \tag{6.36}$$

Figure 6.18 Voltage at point L.

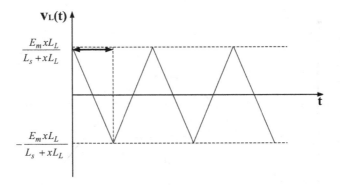

FIGURE 6.18 Voltage at point L.

With what is mentioned above, the voltage across the breaker will be written, as follows:

$$V_{CB}(t) = V_s(t) - V_L(t) \qquad (6.37)$$

Figure 6.19 shows the voltage across the breaker. In practice, the amplitude of the sawtooth wave on sine wave will decrease as the time increases considering its high frequency, and eventually, there will be a sine wave with the angular frequency of ω_1.

6.6 MAGNETIZING CURRENT CHOPPING

Consider Figure 6.20. In the cases presented in this figure, the magnetizing current of the transformer, reactor, or motor should be interrupted with a breaker, which has a considerable rating (Figure 6.21).

The equivalent capacitance of the winding of the transformer (reactor or motor) to the ground is modeled by C. Its range can be the same as the one presented in Figure 6.16. The inductor L energy before de-energization is:

$$E_L = \frac{1}{2} L i_L^2 = \frac{1}{2} C V_C^2 \qquad (6.38)$$

As given in Equation (6.38), this energy can be exchanged with the capacitor C. Then, we can conclude from the above equation that voltage across the capacitor can be:

$$V_C = \sqrt{\frac{L}{C}} i_L \qquad (6.39)$$

The maximum of i_L is $\dfrac{V_m}{2\pi f L}$: therefore, the maximum of the capacitor voltage can be:

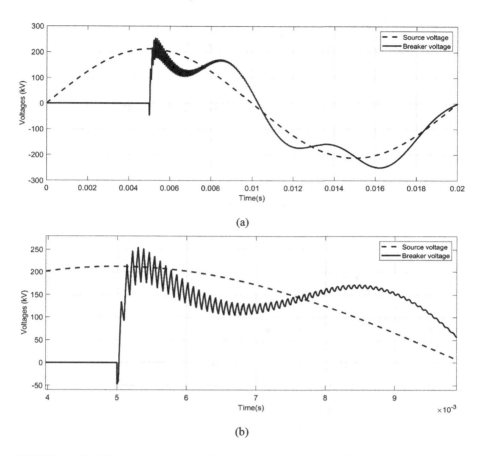

FIGURE 6.19 Voltage across breaker $V_{CB}(t)$: (a) zoomed out and (b) zoomed in.

$$V_C = \sqrt{\frac{L}{C}} \times \frac{V_m}{2\pi fL} \qquad (6.40)$$

Defining $f_0 \triangleq \dfrac{1}{2\pi\sqrt{LC'}}$ Equation (6.40) will become:

$$V_C = \frac{f_0}{f} V_m \qquad (6.41)$$

Figure 6.22 shows the voltage of points A and B (two nodes across the breaker). Figure 6.23a shows the breaker voltage. In this figure, the graph for breaker contacts insulation withstand is shown as well. This curve is similar to the one shown in Figure 6.4. If the withstand curve (dotted curve) is higher than the voltage across the breaker (solid curve), then, we will have a successful operation of the breaker without

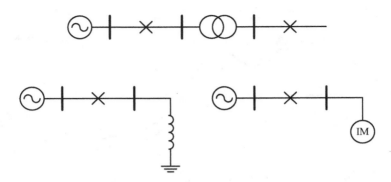

FIGURE 6.20   Magnetizing current interruption, possible practical cases.

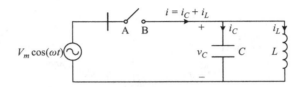

FIGURE 6.21   Equivalent circuit for magnetizing current interruption.

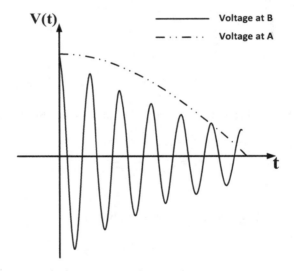

FIGURE 6.22   Voltage at points A and B.

any arc in the breaker. As can be seen in Figure 6.23a, we do not have this condition. When the voltage across the breaker reaches the limit for insulation withstand, the air gap cannot tolerate the voltage and an arc appears across the breaker and voltage across the breaker becomes zero. As the arc extinguishes, the voltage across breaker starts rising. This time, the voltage reaches the insulation withstand at a higher level.

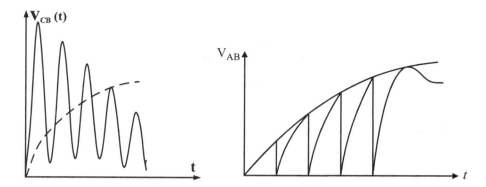

FIGURE 6.23 Voltage across the breaker along with insulation withstand curve (a) without considering any arc and (b) considering arc effect on breaker voltage.

This is because the electrodes are placed farther from each other. The arc continues several times until breaker voltage comes below the insulation withstand curve. It must be mentioned that repetition of this phenomenon can damage the breaker.

In old transformers, magnetizing current is between 0.5% and 2% of the nominal current. In newer transformers, this range is from 0.2% to 0.5% of the nominal current. Therefore, the arcing situation is alleviated in newer transformers.

6.7 RESIDUAL FLUX EFFECT ON TRANSFORMER INRUSH CURRENT

Due the nonlinearity existing in B-H curve of transformers, magnetizing current is not a sinusoidal current. In steady-state conditions, current is bounded between $-I_p$ and $+I_p$ and flux boundary is $-\varphi_m$ to $+\varphi_m$ (see Figure 6.24). When the breaker opens, some flux marked with φ_1 remains in the magnetic circuit, which is known as residual flux (φ_1). This is less than φ_R that is shown in Figure 6.24. In re-energization, the flux will change between $\varphi_1 + \varphi_m$ and $\varphi_1 - \varphi_m$. So, in case of having a positive value for φ_1, in one direction, the core may become highly saturated considering the equation $\varphi_1 + \varphi_m$ and a large amount of current will be drawn by the transformer, which is known as inrush current. Also, in the opposite alter, current magnitude that is drawn is less considering the equation $\varphi_1 - \varphi_m$.

Let us consider a no-load lossless transformer equivalent circuit when it is energized as it is shown in Figure 6.25. We assume that the transformer is being energized at the zero cross of the voltage, and the voltage waveform is as follows:

$$v = V_m \sin \omega t \tag{6.42a}$$

Also, we have:

$$v = \frac{d\lambda}{dt} = \frac{d}{dt}(N\Phi_1 + N\Phi_2) \tag{6.42b}$$

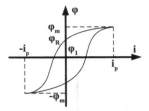

FIGURE 6.24   Transformer hysteresis curve.

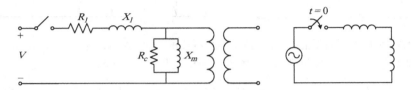

FIGURE 6.25   Transformer equivalent circuit during its energization.

Parameter λ, which is the flux linkage, can be found by the following equation:

$$\lambda - \lambda_0 = \int_0^t V_m \sin(\omega t) dt \qquad (6.43)$$

The above equation will give λ as a function of time. If at $t = 0$, the residual flux is zero ($\lambda_0 = 0$), then we have:

$$\lambda = \frac{V_m}{\omega}(1 - \cos(\omega t)) \qquad (6.44)$$

If at energization instant, there is a residual flux, the linkage flux will be calculated as follows:

$$\lambda = \frac{V_m}{\omega}(1 - \cos(\omega t)) + \lambda_0 \qquad (6.45)$$

Considering the above equations and λ-I characteristic of the transformer core, the inrush current can be plotted as shown in Figure 6.26.

As mentioned above, positive polarity of the residual flux can help the transformer getting saturated, and its opposite polarity can reduce the inrush current. It must be noted that the above equations show that inrush current is dependent on residual flux as well as the energization time (random variables). Therefore, it can be considered as a stochastic phenomenon and can be studied by, for example, Monte Carlo simulation (MCS) method.

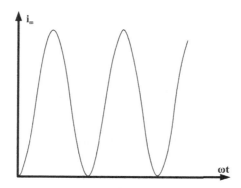

FIGURE 6.26   Nonlinear inrush current of transformer due to residual flux.

In transformer energization, it usually takes 10 cycles or more to reach the steady-state condition. The mentioned time constant in cycle can vary based on self-inductance of the core and nonlinear core losses. The number 10 cycles is an engineering approximation from transients studies. For example, a three-phase 1000 kVA, 13.8 kV transformer, has the nominal current of 42A and magnetizing current of 2.1 A. Inrush current for this transformer is about 150 A. Sometimes, protection system assigned to the transformer at star, cannot differentiate the inrush current from fault current and gives the breaking command. The criteria for differentiating the inrush current from fault current is the second-order harmonic content in inrush current. Protection systems for transformers have the capability of identifying inrush current using the second-order harmonic determination.

6.8   CAPACITIVE CURRENT INTERRUPTION

Capacitive current interruptions and their related transients can be found in disconnecting capacitor banks, disconnecting charged cable, or a long transmission line and disconnection of a series filter. Figure 6.27a shows a circuit, which can model the capacitive current interruption. Breaker opens at $t = t_0$. Figure 6.27b shows voltages at points A and B after the breaker operation. It is assumed that the capacitor has been fully charged before opening the breaker and voltage across the capacitor is equal to the source voltage at the breaking instant.

At $t = 0$, the voltage at the breaker reaches $2V_m$. Assume that at this instant, the breaker contacts airgap cannot withstand this voltage difference, and we have an arc between the contacts. Therefore, the following equation can be written for the resulting circuit.

$$V_m = V_C + V_L = L\frac{di}{dt} + V_C(0) + \frac{1}{C}\int i\,dt \tag{6.46}$$

$$L\frac{di}{dt} + \frac{1}{C}\int i\,dt = V_m - V_C(0) \tag{6.47}$$

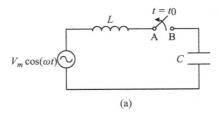

FIGURE 6.27A Capacitive current interruption model.

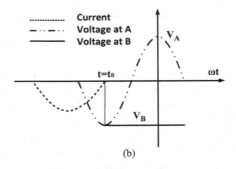

FIGURE 6.27B Current and voltage of points A and B.

In Laplace domain, we will have:

$$LSI(S) - Li(0) + \frac{I(S)}{CS} = \frac{V_m - V_c(0)}{S} \tag{6.48}$$

$$I(S) = \frac{V_m - V_C(0)}{L} \times \frac{1}{S^2 + \omega_0^2} \tag{6.49}$$

In the above equation, $\omega_0{}^2 = \frac{1}{LC}$. Using the Laplace inverse, we will have:

$$i(t) = [V_m - V_C(0)] \times \sqrt{\frac{L}{C}} \sin(\omega_0 t) \tag{6.50}$$

$$v_C(t) = V_C(0) + \frac{1}{C}\int_0^t [V_m - V_C(0)] \times \sqrt{\frac{L}{C}} \sin(\omega_0 t) dt \tag{6.51}$$

Capacitor voltage at the opening is $-V_m$. Therefore,

$$v_C(t) = -V_m + 2V_m[1 - \cos(\omega_0 t)] \tag{6.52}$$

According to the latest equation, the maximum of the capacitor voltage can reach to $-3V_m$. The capacitor voltage is plotted in Figure 6.28. Figure 6.29 shows $V_A - V_B$

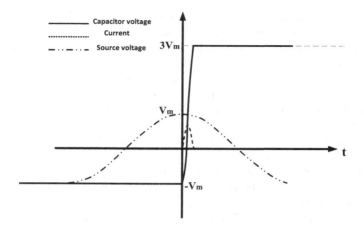

FIGURE 6.28 Breaker voltage and breaker insulation curve.

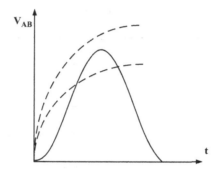

FIGURE 6.29 Breaker voltage after $t = 0$ and insulation withstand curves of two types of circuit breakers.

after breaker opening at $t = t_0$. The insulation withstand curve of oil and SF6 breakers are also shown here. The oil breaker insulation withstand curve is the lower one. It is obvious that for the lower curve that there is an intersection with $V_A - V_B$ curve, which means an arc and unsuccessful operation of the breaker. Therefore, in order to interrupt capacitive currents, SF6 or air blast circuit breakers are used in practice. In Figure 6.29, the upper curve represents a typical case of their insulation withstand curve.

6.9 DISCONNECTORS OPENING

In power system networks, breakers are used to interrupt the electric arc that appears due to fault or load current opening. There is another type of switch that are called disconnectors. They are used to mechanically isolate the circuit or device, which is under voltage but does not have any current. These elements do not have the mechanisms for opening/breaking currents. If by mistake, it opens first in a power circuit, the arc will not be extinguished and that will cause damage to the disconnectors and maybe the equipment. However, as a typical case, we assume in Figure 6.30 that first, the circuit breaker opens, and then after a few seconds, disconnector will mechanically isolate the transformer, which is located after the main circuit breaker and not shown in this figure.

After the main circuit breaker opening, the network illustrated in Figure 6.30 can be presented as in Figure 6.31 for circuit analysis. In this figure, the source side has been modeled by a voltage source behind a T-form LCL circuit. The LCL element amounts depend on the system features. As we have discussed in previous chapters, the capacitances are the dominant elements at the start of a transient phenomenon. This is valid in this case, as well.

In Figure 6.31, C_2 is the capacitive charge model of the disconnector, a short busbar, and/or a transformer. The source side is modeled by the voltage u. Now, the above circuit can be analyzed in two different conditions.

The first condition is when $C_1 \gg C_2$: In this condition, the circuit current will be capacitive with leading phase shift as shown in Figure 6.32 by a dotted curve. At the

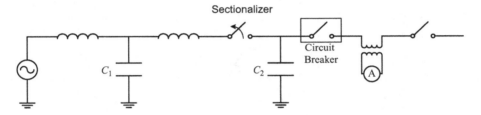

FIGURE 6.30 Power network equivalent circuit with breaker and disconnector.

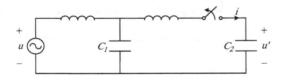

FIGURE 6.31 Circuit equivalent after transformer isolation.

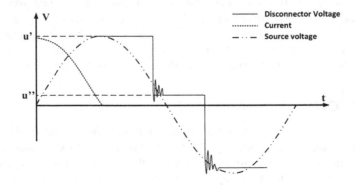

FIGURE 6.32 Voltage across disconnector when $C_1 \gg C_2$.

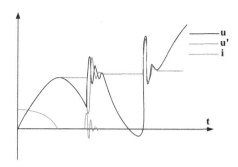

FIGURE 6.33 Voltage across disconnector when $C_1 \ll C_2$.

opening time of the disconnector, when the current is zero and the source side voltage is maximum, voltage u' remains constant across the C_2. On the left side of the disconnector, the voltage is approximately equal to the source voltage. As time passes, the voltage difference across the disconnector gets larger and causes an arc between two sides of the disconnector. At this time, until the arc is extinguished again, the voltage across the C_2 is the same as the source voltage, but with a damped oscillation due to circuit inductances and losses (not shown in the figure). When the arc is extinguished, the voltage at C_2 remains constant at another value of voltage as shown in Figure 6.32. On the left side of the disconnector again, the voltage is equal to the source side voltage. As time passes, the voltage difference across the disconnector gets larger and causes a second arc. The arcing can be repeated a few times in a period. As time passes, the distance between disconnector contacts gets larger. Therefore, the disconnector contacts insulation withstand can tolerate the voltage deference and the circuit will completely be disconnected.

The second condition is when $C_1 \ll C_2$: In this condition, opening the disconnector will cause the u' voltage to remain across the C_2, as shown in Figure 6.33. As the time passes, the airgap insulation strength gets weaker and the arc will appear. The arc current is shown in this figure, as well. As it can be seen up to this point, first condition and second condition are providing the same outcomes. At this point, because C_2 is larger than C_1, the voltage u' stays fixed. Capacitor C_2 charges the capacitor C_1. Therefore, there will be a voltage shift for the source side. Again, the procedure can be repeated, and arcing can happen a few times in a period. It is obvious that in this case, we have a weak source.

6.10 PSCAD EXAMPLES

6.10.1 Transient Recovery Voltages

In this part, some of the examples discussed previously are modeled in PSCAD software and results are discussed. The first is assessing the overvoltage that appeared across a breaker when opening a system fault. Figure 6.34 illustrates the system modeled in PSCAD. The source is considered a cosine wave with an RMS value of 150 kV at 50 Hz frequency. The circuit inductor is 0.3 H and the circuit capacitor is

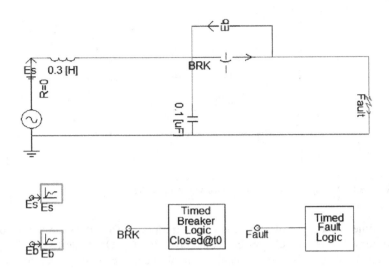

FIGURE 6.34 TRV simulation system.

0.1 uF. Fault happens at $t = 0.02$ s and breaker receives the opening command almost instantaneously. Breaker opening resistance is considered 10,000 Ω. Figure 6.35 illustrates the system voltage and voltage across the breaker terminals.

6.10.2 EFFECT OF EXISTENCE OF TWO NATURAL FREQUENCIES ON TRANSIENT RECOVERY VOLTAGE

Figure 6.36 illustrates a PSCAD representation of a power circuit with two natural frequencies. Source is a cosine wave with an RMS value of 150 kV at 50 Hz. Figure 6.36 contains the values considered for capacitor and inductor. Breaker opens at $t = 0.02$ s. The breaker opening resistance is 10,000 Ω. The 1 Ω resistance has been considered in order to clearly present damping.

Figure 6.37 presents voltage source signal and voltage across the breaker. As it can be seen in this figure, there are two frequencies in the voltage signal. Changing the values of inductors and capacitors in the system will change the shape of the voltage across the breaker.

6.10.3 MAGNETIZING CURRENT CHOPPING

In this part, we are assessing the magnetizing current that is discounted due to a breaker operation. Figure 6.38 shows the PSCAD implementation of this study. The source is a cosine wave of 150 kV RMS at 50 Hz frequency. The inductor and capacitor values are labeled in the figure. Breaker opens at $t = 0.02$ s. Opening resistance of the breaker is 20,000 Ω. Figure 6.39a shows the source voltage and the capacitor voltage. Figure 6.39b shows voltage across the breaker.

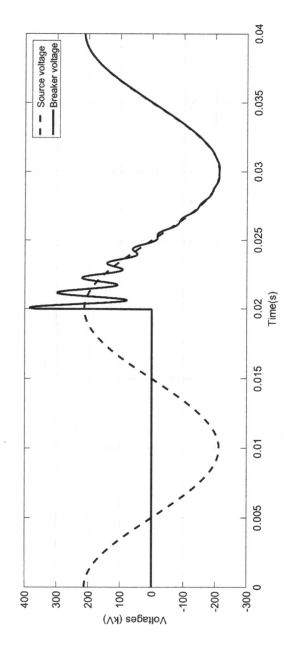

FIGURE 6.35   TRV simulation results.

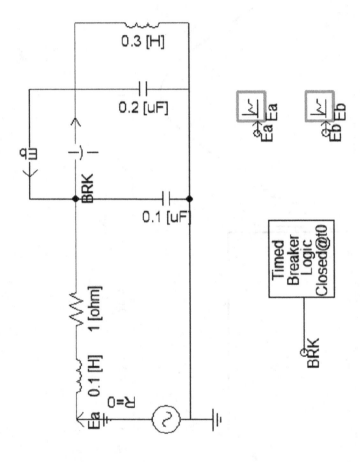

FIGURE 6.36 Simulation system of TRV considering two natural frequencies.

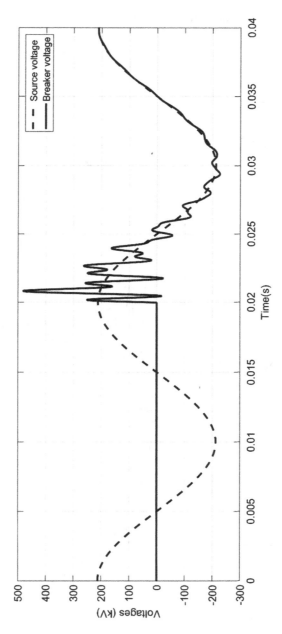

FIGURE 6.37 Simulation results of TRV considering two natural frequencies.

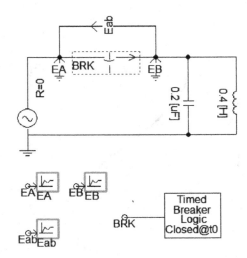

FIGURE 6.38 Simulation system of magnetizing current chopping.

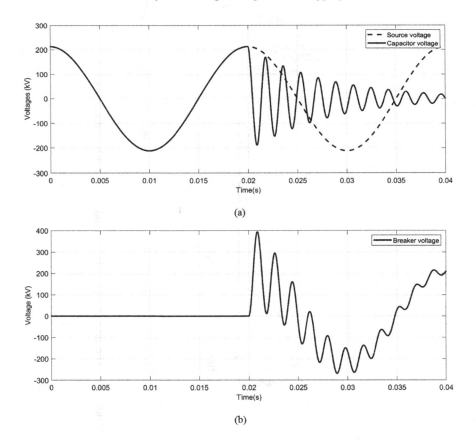

FIGURE 6.39 Simulation results of (a) source and L-C voltages and (b) breaker voltage.

Problems

6.1 In the following circuit, breaker opens at $t = 0.005$.
Using PSCAD modeling, provide the time signal of the voltage across the breaker.

What are the similarities and differences between the obtained figure and Figure 6.19?

6.2 In the following circuit, the RRRV is larger than the accepted value by standards. In order to decrease this value by 20%, a power capacitor has been added in parallel with the breaker. Determine the amount of this capacitor as a function C shown in the figure.

6.3 In the following circuit, the breaker CB2 is closed, when CB1 operates and open the circuit. In the first condition, we assume CB3 is open, and in the second condition, we assume CB3 is closed. Which condition does cause the largest TRV? Explain why.

6.4 In the illustrated figure at $t = 0$, a no-load transformer is isolated from the circuit.

a. Using the current injection method, find the voltage across the breaker CB (ignore the damping effects in the circuit).

b. After drawing the voltage across CB, draw the insulation withstand diagram of the breaker for the condition of no arcing in the breaker.

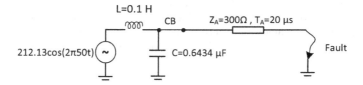

FIGURE P6.1   Fault applied at the end of a transmission line.

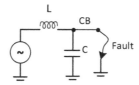

FIGURE P6.2   Fault interruption circuit.

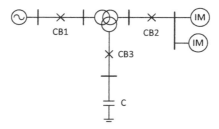

FIGURE P6.3   – Grid to analyze TRV scenarios.

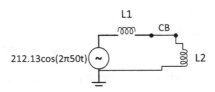

FIGURE P6.4   No-load transformer opening.

BIBLIOGRAPHY

1. C. L. Wadhwa, *"Electrical Power Systems"*, *New Academic Science*, 2011, 900 pages.
2. David F. Peelo, *Current Interruption Transients Calculation*, John Wiley & Sons, 2020.
3. N. H. Malik, A. A. Al-Arainy, and M. I. Qureshi, *Electrical Insulation in Power Systems*, The United States of America: Taylor & Francis Group, 1998.
4. Qingmin Li, Hongshun Liu, Jie Lou, and Liang Zou, "Impact Research of Inductive FCL on the Rate of Rise of Recovery Voltage With Circuit Breakers", *IEEE Transactions on Power Delivery*, Vol. 23, No. 4, October 2008, pp. 1978–1985.
5. P. I. Obi, M. C. Emeghara, and Osita Oputa, "Dynamic modeling of circuit breaker transient recovery voltage", *3rd International Conference on Electro-Technology for National Development (NIGERCON)*, February 2018, Owerri, Nigeria.
6. Hiroki Ito, Hiroki Kajino, Yoshibumi Yamagata, Kenji Kamei, Toshiki Idemaru, and Hiromu Shinkai, "Study on Transient Recovery Voltages for Transformer-Limited Faults", *IEEE Transactions on Power Delivery*, Vol. 29, No. 5, October 2014, pp. 2375–2384.

7 Traveling Wave Influence on Power Transformers Windings

This chapter discusses traveling waves in transformer windings. Here, we are considering a transformer model that suites transient situations as opposed to steady conditions. Concepts discussed in this chapter with minor changes are applicable for reactors and HV motor windings as well. We need to make the following assumptions when dealing with traveling waves in transformers:

1. We just consider one coil. This is because lightning hits mostly one of the phases of a transformer. Therefore, analyzing one phase will be sufficient. In switchings, as discussed in previous chapters, all three phases of a circuit breaker do not close all at the same time. Therefore, the traveling wave reaches one of the phases faster.
2. The coil is coreless. In transients, we are dealing with really high frequencies in the range of kHz and higher. In this condition, flux does not get a chance to penetrate into the core of a power transformer. Actually, the core presents a behavior very similar to skin effects in conductors. The core outer surface will be modeled as an equipotential surface. In other words, air core will be an appropriate model for the transients in frequency ranges around kHz and higher.
3. In order to reduce the leakage flux, the core should fill the space inside the cylinder form winding. In practice, the solution is a step-lapped core. Therefore, the winding surrounds the semi-cylindrical core, and it is possible to consider a circle for the cross-sectional area of the core.

Later on in this chapter, we will see that the transient model of a transmission line is a special case of a general model, which is valid for transformer winding and will be discussed in the following sections.

7.1 WINDING MODELING WITHOUT MUTUAL INDUCTANCES

Figure 7.1 shows a winding in per unit length. In this figure, C', K', and L' are capacitive charge to the ground in per unit length (Farad/meter), parallel capacitive charge per unit length (Farad.meter), and inductance in per unit length (Henry/meter), respectively.

DOI: 10.1201/9781003255130-7

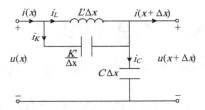

FIGURE 7.1 Winding model in per unit length.

There is current loss in per unit length, which can be written as follows:

$$\begin{cases} i(x) = i_L + i_K \\ i(x + \Delta x) = i_L + i_K - i_C \end{cases} \tag{7.1}$$

From Equation (7.1), we can conclude:

$$\Delta i(x) = i(x + \Delta x) - i(x) = -i_C \tag{7.2}$$

Considering $i_C = C'\Delta x \dfrac{\partial u(x + \Delta x)}{\partial t}$ and combining that with Equation (7.2), we will get:

$$-\frac{\Delta i(x)}{\Delta x} = C'\frac{\partial u(x + \Delta x)}{\partial t} \tag{7.3}$$

In Equation (7.3), if Δx tends to zero, using the definition of derivatives, we will have:

$$-\frac{\partial i(x)}{\partial x} = C'\frac{\partial u(x)}{\partial t} \tag{7.4}$$

For the voltage drop across the parallel capacitor–inductor combination, we will have:

$$\Delta u(x) = u(x + \Delta x) - u(x) \tag{7.5}$$

Using the above voltage drop equation, and also, the relationship between voltage and current of capacitor, the following equation can be written:

$$i_K = \frac{K'}{\Delta x}\frac{\partial}{\partial t}(u(x) - u(x + \Delta x)) \tag{7.6}$$

If Δx tends to zero, using the definition of derivatives, we will have:

$$i_K = -K'\frac{\partial^2 u}{\partial x \partial t} \tag{7.7}$$

Let us consider the equations for voltage drop across the inductor:

$$u(x) - u(x + \Delta x) = L' \Delta x \frac{\partial i_L}{\partial t} \tag{7.8}$$

Again, if Δx tends to zero, using the definition of derivatives, we will have:

$$\frac{\partial u}{\partial x} = -L' \frac{\partial i_L}{\partial t} \tag{7.9}$$

Now, let us take partial derivative of Equation (7.1) with respect to x:

$$i(x) = i_L + i_K \implies \frac{\partial i}{\partial x} = \frac{\partial i_L}{\partial x} + \frac{\partial i_K}{\partial x} \tag{7.10}$$

Implementing Equations (7.4) and (7.7) in Equation (7.10), we will have:

$$\frac{\partial i_L}{\partial x} = -C' \frac{\partial u}{\partial t} + K' \frac{\partial^3 u}{\partial x^2 \partial t} \tag{7.11}$$

Now, we take a partial derivative from both sides of Equation (7.11) and multiply both sides by L'.

$$L' \frac{\partial^2 i_L}{\partial x \partial t} = -L'C' \frac{\partial^2 u}{\partial t^2} + L'K' \frac{\partial^4 u}{\partial x^2 \partial t^2} \tag{7.12}$$

Now taking the partial derivative of Equation (7.9) with respect to x and placing it in Equation (7.12), we will have:

$$\frac{\partial^2 u}{\partial x^2} - L'C' \frac{\partial^2 u}{\partial t^2} + L'K' \frac{\partial^4 u}{\partial x^2 \partial t^2} = 0 \tag{7.13}$$

7.1.1 VOLTAGE INITIAL DISTRIBUTION

From circuit theory, we know that the inductor acts as an open circuit and capacitor acts as a short circuit in initial conditions. Therefore:, we can consider the winding model as shown in Figure 7.2. For solving the equations, we consider two assumptions.

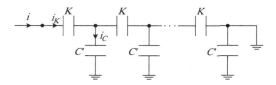

FIGURE 7.2 Equivalent circuit for initial condition.

First, input is a step function (we have explained the reason why using step function as the input in Chapter 2). This assumption can be written as $u(0,0) = U_0$. Second, the end of the coil is grounded. Figure 7.2 shows the equivalent circuit. In these conditions, i_L in Equation (7.1) is zero. Taking partial derivative from Equation (7.1) with respect to position and considering $i_L = 0$, we have:

$$-\frac{\partial i(x)}{\partial x} = C' \frac{\partial u(x)}{\partial t} \tag{7.14}$$

Using Equations (7.7) and (7.14), we will have:

$$\frac{C'}{K'} \frac{\partial u_0(x)}{\partial t} = \frac{\partial^3 u_0(x)}{\partial^2 x \partial t} \tag{7.15}$$

We can find Equation (7.15) directly from Equation (7.13). Using Equation (7.13), we have:

$$L' = \frac{\dfrac{\partial^2 u}{\partial x^2}}{C' \dfrac{\partial^2 u}{\partial t^2} - K' \dfrac{\partial^4 u}{\partial x^2 \partial t^2}} \tag{7.16}$$

Considering open circuit condition for the inductor initially and an infinite value for L' in Equation (7.16), we set the denominator to zero and the result will be the same as Equation (7.15). The solution for Equation (7.15) is in the form of the following equation:

$$u_0(x) = A e^{\gamma x} + B e^{-\gamma x} \tag{7.17}$$

Which $\gamma = \sqrt{C'/K'}$

In order to find the coefficient A and B, we consider borderline conditions. At the beginning of the winding ($x = 0$), voltage is U_0, and at the end of the winding ($x = l$), voltage is zero. Therefore, A and B can be found as:

$$\begin{cases} A = U_0 \dfrac{e^{-\gamma l}}{e^{-\gamma l} - e^{\gamma l}} \\[4mm] B = U_0 \dfrac{-e^{\gamma l}}{e^{-\gamma l} - e^{\gamma l}} \end{cases} \tag{7.18}$$

The solution will be:

$$u_0(x) = U_0 \frac{e^{-\gamma(l-x)} - e^{\gamma(l-x)}}{e^{-\gamma l} - e^{\gamma l}} \tag{7.19}$$

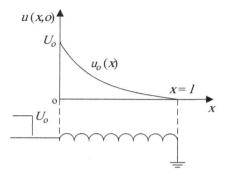

FIGURE 7.3 Initial voltage distribution versus position.

This equation is the definition of hyperbolic-sine:

$$u_0(x) = U_0 \frac{\sin h(\gamma(l-x))}{\sin h(\gamma l)} \tag{7.20}$$

Figure 7.3 shows the diagram of the voltage distribution with respect to x. As it can be seen, the voltage at the beginning of the winding is the highest and its value decreases with Equations (7.19) or (7.20) and it reaches zero at the end of the winding (end of the coil is grounded).

Using Equation (7.20), the voltage gradient in initial conditions can be found from Equation (7.21). This gradient can be considered as the potential difference between two adjacent discs of an HV winding, which is important for the insulation design of windings.

$$\frac{\partial u_0(x)}{\partial x} = -U_0 \frac{\gamma \cos h(\gamma(l-x))}{\sin h(\gamma l)} \tag{7.21}$$

The maximum of the voltage gradient at $(x = 0)$ will be calculated as:

$$\max\left(\left.\frac{\partial u_0(x)}{\partial x}\right|_{x=0}\right) = U_0 \gamma \cot h(\gamma l) \tag{7.22}$$

This tells us that the maximum of the voltage gradient at the beginning of the winding appears at the initial instants when the step wave is applied to the winding. Figure 7.4 shows the voltage diagram with respect to the position in different γs. As it can be seen, as $\dfrac{C'}{K'}$ ratio grows (C' transversal capacitor and K' is the longitudinal capacitor in Figure 7.1), the slope of the voltage drop with respect to position becomes sharper. In other words, when the step wave hits the beginning of the winding, the tension is maximum at that point, and in designing transformer winding insulation, considerations need to be made in order to increase the value of K'.

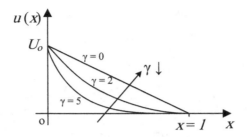

FIGURE 7.4   Voltage with respect to position; effect of parameter γ.

7.1.2   EQUIVALENT CAPACITANCE OF THE WINDING AT $T = 0$

We consider C_0 as the equivalent capacitor of the chain of C' and K' shown in Figure 7.2. The equivalent capacitor C_0 draws the same amount of current as the other capacitors. As stated before, the inductor is considered an open circuit at $t = 0$. Therefore, the current i_k goes through the existing capacitive chain. Using Equation (7.7) and replacing $u_0(x)$ instead of $u(x)$ at initial conditions, we will have:

$$i_K = -K' \frac{\partial}{\partial t}\left(\frac{\partial u_0(x)}{\partial x}\right) \tag{7.23}$$

On the contrary, we use the relationship between current and voltage of the capacitor as the following:

$$i_K = C_0 \frac{\partial u_0(x)}{\partial t} \tag{7.24}$$

Considering Equations (7.23) and (7.24) as two equal equations:

$$C_0 u_0(x)\big|_{x=0} = -K' \frac{\partial u_0(x)}{\partial x}\bigg|_{x=0} \tag{7.25}$$

Now, considering Equation (7.22), we can rewrite the latest equation as:

$$C_0 U_0 = K' U_0 \gamma \coth(\gamma l) \tag{7.26}$$

Therefore, the equivalent capacitor of the capacitive chain will be:

$$C_0 = \sqrt{C'K'} \coth(\gamma l) \tag{7.27}$$

Considering the fact that in practice, $\gamma l > 3$, we can consider the approximation of $\cosh(\gamma l) \cong 1$. Therefore, the equivalent capacity will be $C_0 = \sqrt{C'K'}$, which is the geometric mean of the two capacitors C' and K'.

7.1.3 FINAL DISTRIBUTION

In the final distribution of the step wave $u_e(x)$, according to circuit theory foundations, the effects of the capacitors will become negligible ($C' = K' = 0$). Therefore, we have:

$$\frac{\partial^2 u_e(x)}{\partial x^2} = 0 \qquad (7.28)$$

Solving the above differential equation, the final voltage will look like:

$$u_e(x) = Ax + B \qquad (7.29)$$

In order to find the coefficients A and B, we use borderline conditions:

$$\begin{cases} x = 0 \Rightarrow u_e(0) = U_0 = B \\ x = l \Rightarrow u_e(l) = 0 = Al + B \end{cases}$$

Therefore, the final voltage will be written as:

$$u_e(x) = U_0\left(1 - \frac{x}{l}\right) \qquad (7.30)$$

Figure 7.5 shows the final distribution of the voltage, that is, $u_e(x)$.

7.1.4 SOLUTION FOR MAIN PARTIAL DIFFERENTIAL EQUATION

In this section, we put some effort into finding the direct solution for Equation (7.13), which is repeated below.

$$\frac{\partial^2 u}{\partial x^2} - L'C'\frac{\partial^2 u}{\partial t^2} + L'K'\frac{\partial^4 u}{\partial x^2 \partial t^2} = 0$$

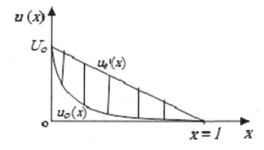

FIGURE 7.5 Final distribution of the voltage across the winding.

Considering the method of separation of variables, the solution for the equation $u(x, t)$ will be shown as the product of two different functions. One of the functions is solely dependent on x and the other on t meaning:

$$(x,t) = f(x)g(t) \tag{7.31}$$

Implementing Equation (7.31) in the partial derivative differential Equation (7.13), we will have:

$$f''g - L'C'g''f + L'K'f''g'' = 0 \tag{7.32}$$

Equation (7.32) is rewritten as:

$$\frac{f''}{f} = \frac{L'C'g''}{g + L'K'g''} \tag{7.33}$$

In this equation, the left-hand side of the equation is a function of x and the right-hand side is the function of t; therefore, in order to solve the equation, we can assume $\frac{f''}{f} = -\alpha^2$. Therefore, function f will be:

$$f(x) = A\cos(\alpha x) + B\sin(\alpha x) \tag{7.34}$$

In order to find function g, we follow the same methodology as above:

$$\frac{L'C'g''}{g + L'K'g''} = -\alpha^2 \tag{7.35}$$

Equation (7.35) is reordered as $g'' + \omega^2 g = 0$, which:

$$\omega = \frac{\alpha}{\sqrt{L'C' + L'K'\alpha^2}} \tag{7.36}$$

Therefore, the solution will be:

$$g(t) = C\cos(\omega t) + D\sin(\omega t) \tag{7.37}$$

Overall, the solution of the differential equation will be presented as:

$$u(x,t) = (A\cos(\alpha x) + B\sin(\alpha x)) \times (C\cos(\omega t) + D\sin(\omega t)) + U_0\left(1 - \frac{x}{l}\right) \tag{7.38}$$

Now, with borderline conditions as having U_0 at the beginning of the winding:

$$U_0 = A \times (C\cos\omega t + D\sin\omega t) + U_0$$

Therefore, A is calculated to be zero, knowing the fact that $(C\cos(\omega t) + D\sin(\omega t))$ is not zero.

At the end of the winding, we have $u(x = l, t) = 0$. Therefore:

$$0 = B\sin(\alpha x) \times (C\cos(\omega t) + D\sin(\omega t))$$

As we know $(C\cos(\omega t) + D\sin(\omega t))$ is not zero, therefore, $B\sin(\alpha x)$ is zero. We conclude $\alpha_n = \dfrac{n\pi}{l}$ and thus, voltage with respect to time and position will be written as:

$$u_n(x,t) = \sin\left(\frac{n\pi}{l}x\right) \times (C\cos(\omega t) + D\sin(\omega t)) + U_0\left(1 - \frac{x}{l}\right) \tag{7.39}$$

Considering the fact that Equation (7.39) is valid for all n-s, therefore:

$$u(x,t) = \sum_n (C_n\cos(\omega_n t) + D_n\sin(\omega_n t))\sin\left(\frac{n\pi}{l}x\right) + U_0\left(1 - \frac{x}{l}\right)$$

Let us assume $\left.\dfrac{\partial u(x,t)}{\partial t}\right|_{t=0} = 0$. With this assumption, the parameter D will be zero, and voltage as a function of time and position will be written as:

$$u(x,t) = \sum_n C_n\sin\left(\frac{n\pi}{l}x\right)\cos(\omega_n t) + U_0\left(1 - \frac{x}{l}\right) \tag{7.41}$$

At $t = 0$, we have:

$$u(x, t = 0) = u_0(x) = \sum_n C_n\sin\left(\frac{n\pi}{l}x\right) + u_e(x) \tag{7.42}$$

We can conclude from Equation (7.42):

$$u_0(x) - u_e(x) = \sum_n C_n\sin\left(\frac{n\pi}{l}x\right) \tag{7.43}$$

In order to find the coefficients in the above equation, we find the difference between the initial distribution and the final distribution. With the assumption that the voltage wave is a periodic one, we can write the Fourier series, and determine its coefficients, as follows:

$$C_n = -\frac{2}{n\pi}\frac{\gamma^2 l^2}{(n\pi)^2 + \gamma^2 l^2} \tag{7.44}$$

According to Equation (7.43), which is for the lossless condition and neglecting the mutual coupling, we can say that at $t = 0$, the difference of initial and final voltage distributions, that is, $u_0(x) - u_e(x)$, is decomposed by Fourier analysis in length l. It is interesting that the sum of these waves (terms of Fourie series) is a traveling wave penetrated in the winding. This point will be discussed in the following subsection.

7.1.5 TRAVELING WAVE RESPONSE JUSTIFICATION

In order to analyze why the obtained voltage in Equation (7.41) is a traveling wave, we rewrite this equation as:

$$
\begin{aligned}
u(x,t) &= \sum_n C_n \sin(\alpha_n x)\cos(\omega_n t) + U_0\left(1 - \frac{x}{l}\right) \\
&= \sum_n \frac{1}{2} C_n[\sin(\alpha_n x + \omega_n t) + \sin(\alpha_n x - \omega_n t)] + U_0\left(1 - \frac{x}{l}\right)
\end{aligned}
\tag{7.45}
$$

Considering the fact that $\omega_n = 2\pi f_n = 2\pi\left(\dfrac{\vartheta_n}{\lambda_n}\right)$, we can write the sine argument in Equation (7.45) as $\left(\alpha_n x + \dfrac{2\pi}{\lambda_n}\vartheta_n t\right)$. Therefore, Equation (7.45) can be presented as $f(x + \vartheta_n t)$ and $f(x - \vartheta_n t)$, which represent traveling waves in both directions of the x-axis. Speed of the wave propagation is $\vartheta_n = \omega_n \big/ \alpha_n$ and $\omega_n = \dfrac{\alpha_n}{\sqrt{L'C' + L'K'\alpha_n^2}}$, therefore, we can write:

$$
\vartheta_n = \frac{\omega_n}{\alpha_n} = \frac{1}{\sqrt{L'C' + L'K'\alpha_n^2}} = \frac{1}{\sqrt{L'C'\left(1 + \dfrac{\alpha_n^2}{\gamma^2}\right)}}
\tag{7.46}
$$

If $n = 0$, then $\vartheta_0 = 1\big/\sqrt{L'C'}$. Using this equation, we can write the speed of the wave as:

$$
\vartheta_n = \frac{\vartheta_0}{1 + \left(\dfrac{\alpha_n}{\gamma}\right)^2} = \frac{\vartheta_0}{1 + \left(\dfrac{n\pi}{\gamma l}\right)^2}
\tag{7.47}
$$

Analyzing Equation (7.47), we conclude if $K' = 0$ (i.e., ($\gamma \to \infty$)), the wave speed will be: $\vartheta_n = 1\big/\sqrt{L'C'}$ which is a constant value. This speed is the same as the speed found for the lossless line in Chapter 2. The assumption $K' = 0$ means that there is no longitudinal capacitance in Figure 7.1, and therefore, this figure changes to the model of the lossless line.

As mentioned before, according to Equation (7.43), at $t = 0$, $u_0(x) - u_e(x)$ is decomposed by Fourier analysis in length l. Each element of the obtained Fourier series has a particular speed of propagation in the winding. Therefore, the wave does not preserve its shape during the time of propagation. In transmission lines if we ignore G and R, the traveling wave preserves its shape. This shows how complicated is to analyze the transient condition in a winding, compared to a transmission line. Now considering Equation (7.47), if n tends to infinity, α_n also tends to infinity and ϑ_n tends to zero. Considering the natural frequency of a turn of a winding as:

$$\omega_{\text{crit}} = \lim_{n \to \infty} \omega_n = \frac{1}{\sqrt{L'K'}} \tag{7.48}$$

Based on this model, we can conclude that for all waves with $\omega_{\text{crit}} < \omega$, they will not enter the winding. In order to improve the modeling, we must consider mutual coupling and winding losses.

7.2 WINDING MODELING WITH MUTUAL COUPLING AND RESISTANCES

This analysis can be used in transformers, reactors, and generator windings for transient studies. It is required to note that for transient conditions, the winding has an air core, and the core can be modeled by an equipotential cylinder surrounded by the winding.

7.2.1 WINDING DETAILED MODEL

Let us assume f_{max} is the maximum frequency in the study. This frequency for switching studies can be around 20 kHz or for GIS switching studies; it can be 5 MHz. If v is the speed of wave propagation, therefore, the smallest wavelength in the study will be:

$$\lambda_{\text{min}} = \frac{\vartheta}{f_{\text{max}}} \tag{7.49}$$

If the length of the largest element in the model is at most $0.2\lambda_{\text{min}}$ (i.e., $l_{\text{max}} \le 0.2\lambda_{\text{min}}$), then it is possible to change a distributed model to a lumped model. Therefore, instead of looking for partial differential equations, we look for ordinary differential equations. In order to reach the condition of using ordinary differential equations, we divide the model into n parts/segments with the abovementioned condition. This segmented model, known as the detailed model, is illustrated in Figure 7.6.

One of the main issues of the detailed model, as shown in Figure 7.6, is the parameter calculation or estimation of this model. Also, there are challenges in solving equations in time and frequency domains numerically.

For this model, we need to consider that each node has a minimum of one capacitive, one inductive, and one resistive branch. According to Figure 7.6, we can find out

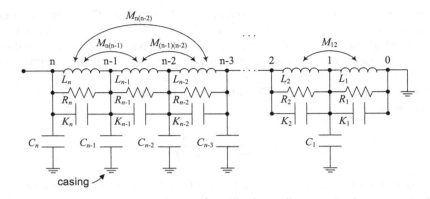

FIGURE 7.6 *n* part model (detailed model).

that there is an inductor connected to each node of the grid, and therefore, voltage equations for the inductors in the grid can be written as follows:

$$
\begin{cases}
u_{L1} = L_1 \dfrac{di_{L1}}{dt} + M_{12} \dfrac{di_{L2}}{dt} + \cdots + M_{1n} \dfrac{di_{Ln}}{dt} \\[2mm]
u_{L2} = M_{21} \dfrac{di_{L1}}{dt} + L_2 \dfrac{di_{L2}}{dt} + \cdots + M_{2n} \dfrac{di_{Ln}}{dt} \\[2mm]
\quad\vdots \\[2mm]
u_{Ln} = M_{n1} \dfrac{di_{L1}}{dt} + M_{n2} \dfrac{di_{L2}}{dt} + \cdots + L_n \dfrac{di_{Ln}}{dt}
\end{cases}
\tag{7.50}
$$

Equation (7.50) can be written in a matrix form as:

$$
U_{bL} = M \dot{i}_{bL}
\tag{7.51}
$$

Which,

$$
U_{bL} = \begin{bmatrix} U_{bL1} = U_{L1} \\ U_{bL2} = U_{L2} \\ \vdots \\ U_{bLn} = U_{Ln} \end{bmatrix}
\tag{7.52}
$$

$$
M = \begin{bmatrix}
L_1 & M_{12} & \cdots & M_{1n} \\
M_{21} & L_2 & \cdots & M_{2n} \\
\vdots & \vdots & \cdots & \vdots \\
M_{n1} & M_{n2} & \cdots & M_{nn}
\end{bmatrix}
\tag{7.53}
$$

$$
\dot{I}_{bL} = \begin{bmatrix} i^{\circ}{}_{L1} \\ i^{\circ}{}_{L2} \\ \vdots \\ i^{\circ}{}_{Ln} \end{bmatrix}
\tag{7.54}
$$

Now, we can write the incidence matrix of the network as:

$$
A = \begin{bmatrix}
1 & -1 & 0 & \cdots & 0 & 0 \\
0 & 1 & -1 & \cdots & 0 & 0 \\
\vdots & \vdots & \ddots & \ddots & \vdots & \vdots \\
\vdots & \vdots & \ddots & \ddots & \vdots & \vdots \\
0 & 0 & 0 & \cdots & 1 & -1 \\
0 & 0 & 0 & \cdots & 0 & 1
\end{bmatrix}
\tag{7.55}
$$

In matrix 7.55, each row presents one node and each column presents one branch. Let us use this matrix in voltage and current equations:

$$
U_{bL} = A^T U \tag{7.56}
$$

Using Equations (7.56) and (7.51), we have:

$$
A^T U = M \dot{i}_{bL} \tag{7.57}
$$

Now multiplying both sides of the equation to M, we have:

$$
M^{-1} A^T U = \dot{i}_{bL} \tag{7.58}
$$

Now multiplying both sides of the last equation to matrix A:

$$
A M^{-1} A^T U = A \dot{i}_{bL} \tag{7.59}
$$

$$
\dot{i}_L = HU \tag{7.60}
$$

The right-hand side of Equation (7.59) is I_L. Matrix H is:

$$
H = A M^{-1} A^T \tag{7.61}
$$

H is a symmetric matrix. We can prove this statement by showing the transpose of it is equal to matrix H.

$$
H^T = (A M^{-1} A^T)^T = (A^T)^T (M^{-1})^T A^T = A(M^T)^{-1} A^T = A M^{-1} A^T = H
$$

In the above proof, we used the fact that two inductors have the same mutual inductances (i.e., $M_{mn} = M_{nm}$). Therefore, the transpose of M is the same as M.

Now, we use Figure 7.7 in order to analyze other important properties of matrix H. Considering each branch, we can calculate the current in each branch as:

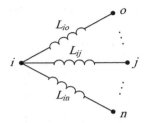

FIGURE 7.7   Equivalent circuit to assess H matrix properties.

The following equation can be written using Figure 7.7.

$$\begin{cases} \dot{I}_{i0} = \dfrac{U_i - U_0}{L_{i1}} \\[2mm] \quad\vdots \\[2mm] \dot{I}_{ij} = \dfrac{U_i - U_j}{L_{ij}} \\[2mm] \quad\vdots \\[2mm] \dot{I}_{in} = \dfrac{U_i - U_n}{L_{in}} \end{cases} \tag{7.62}$$

Considering the above equation, the injected current to point i will be:

$$\dot{I}_i = \dot{I}_{i0} + \cdots + \dot{I}_{ij} + \cdots + \dot{I}_{in} = \frac{1}{L_{i1}}(U_i - U_0) + \cdots + \frac{1}{L_{ij}}(U_i - U_j) + \cdots + \frac{1}{L_{in}}(U_i - U_n) \tag{7.63}$$

Therefore, we have:

$$\dot{I}_i = -\frac{1}{L_{i0}}U_0 + -\frac{1}{L_{i1}}U_1 + \cdots + \left(\sum_{\substack{j=0 \\ j \neq i}}^{n} \frac{1}{L_{ij}} \right) U_i + \cdots + \frac{1}{L_{in}}U_n \tag{7.64}$$

In the latest equation, i is an integer value. If we consider $U_0 = 0$, we will have:

$$\begin{bmatrix} \dot{I}_{L1} \\ \dot{I}_{L2} \\ \vdots \\ \dot{I}_{Ln} \end{bmatrix} = \begin{bmatrix} \displaystyle\sum_{\substack{j=0 \\ j \neq 1}}^{n} \frac{1}{L_{1j}} & -1\!\!\big/\!L_{12} & \cdots & -1\!\!\big/\!L_{1n} \\[4mm] -1\!\!\big/\!L_{21} & \displaystyle\sum_{\substack{j=0 \\ j \neq 2}}^{n} \frac{1}{L_{2j}} & \cdots & -1\!\!\big/\!L_{2n} \\[4mm] \vdots & \vdots & \vdots & \vdots \\[2mm] -1\!\!\big/\!L_{n1} & -1\!\!\big/\!L_{n2} & \cdots & \displaystyle\sum_{\substack{j=0 \\ j \neq n}}^{n} \frac{1}{L_{nj}} \end{bmatrix} \begin{bmatrix} U_1 \\ U_2 \\ \vdots \\ U_n \end{bmatrix} \tag{7.65}$$

We can summarize Equation (7.65) as:

$$\dot{I}_L = \Gamma U \tag{7.66}$$

Comparing Equation (7.66) with (7.60), we see that the two matrices H and Γ are equal. This tells us about the possibility of modeling a network with mutual inductances considering a network without mutual inductances. The only difference is we need to consider an inductor between every two nodes in the network.

For the capacitors of this network, consider Figure 7.8. Considering this at node i, we can write:

$$q_i = C_{i0}(U_i - U_0) + \cdots + C_{ij}(U_i - U_j) + \cdots + C_{in}(U_i - U_n) \tag{7.67}$$

In the latest equation, i is an integer in the series $i = 1,2, \ldots, n$. We can rewrite Equation (7.67) as:

$$q_i = -C_{i0}U_0 - C_{i1}U_1 - \cdots + \left(\sum_{\substack{j=0 \\ j \neq i}}^{n} C_{ij} \right) U_i + \cdots - C_{in}U_n \tag{7.68}$$

Using the equation $q = cv$, we will have:

$$
\begin{bmatrix} q_1 \\ q_2 \\ \vdots \\ q_n \end{bmatrix} =
\begin{bmatrix}
\sum_{\substack{j=0 \\ j \neq 1}}^{n} C_{1j} & -C_{12} & \cdots & -C_{n1} \\
\vdots & \vdots & \vdots & \vdots \\
-C_{n1} & -C_{n2} & \cdots & \sum_{\substack{j=0 \\ j \neq n}}^{n} C_{nj}
\end{bmatrix}
\begin{bmatrix} U_1 \\ U_2 \\ \vdots \\ U_n \end{bmatrix} \tag{7.69}
$$

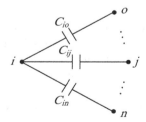

FIGURE 7.8 Auxiliary network to form capacitive matrix.

According to the latest equation in the n-part model, we have:

$$C = \begin{bmatrix} C_1 + K_1 + K_2 & -K_2 & 0 & \cdots & 0 & 0 \\ -K_2 & C_2 + K_2 + K_3 & -K_3 & \cdots & 0 & 0 \\ 0 & -K_3 & C_3 + K_3 + K_4 & \cdots & 0 & 0 \\ \vdots & \vdots & \vdots & \vdots & \vdots & \vdots \\ 0 & 0 & 0 & \cdots & C_n + K_{n-1} + K_n & -K_n \\ 0 & 0 & 0 & \cdots & -K_n & C_n + K_n \end{bmatrix} \quad (7.70)$$

In the same way, for resistances, we have: $I_G = GU$ and $\dot{I}_G = G\dot{U}$.
G matrix is formed as in Equation (7.71):

$$G = \begin{bmatrix} \dfrac{1}{R_1} + \dfrac{1}{R_2} & -\dfrac{1}{R_2} & 0 & \cdots & 0 & 0 \\ -\dfrac{1}{R_2} & \dfrac{1}{R_2} + \dfrac{1}{R_3} & -\dfrac{1}{R_3} & \cdots & 0 & 0 \\ 0 & -\dfrac{1}{R_3} & \dfrac{1}{R_2} + \dfrac{1}{R_3} & 0 & -0 & 0 \\ \vdots & \vdots & \vdots & \vdots & \vdots & \vdots \\ 0 & 0 & 0 & \cdots & \dfrac{1}{R_n} + \dfrac{1}{R_{n-1}} & -\dfrac{1}{R_n} \\ 0 & 0 & 0 & \cdots & -\dfrac{1}{R_n} & \dfrac{1}{R_n} \end{bmatrix} \quad (7.71)$$

The KCL in the detailed model can be written as:

$$I = I_C + I_G + I_L \quad (7.72)$$

Taking derivative with respect to time from both sides of the last equation, we will come up with:

$$\dot{I} = \dot{I}_C + \dot{I}_G + \dot{I}_L \quad (7.73)$$

$$\dot{I} = C\ddot{U} + G\dot{U} + H \quad (7.74)$$

As it can be seen, we have an ordinary second-order differential equation now. Therefore, instead of having the distributed model, we can use the lumped model. The input to this model is the vector I and the output is the node voltage vector U in the time domain. Thus, we can determine the voltage gradient on insulations between two adjacent nodes. The detailed model can be developed and solved by software such as PSCAD and EMTP. Also, core behavior and frequency-dependent losses can be added to this model.

In the end, it must be mentioned that switching in gas insulated substations (GIS) results in very steep wavefronts and fast oscillations. Proper representation of these waves needs the transformer model to be accurate at frequencies often extending far beyond 1 MHz. It is questionable if any of the available white-box, black-box, or gray-box models are suitable for such studies considering limitations in their frequency band. To model the impact of the transformer on the initial steep wave front, a surge capacitance per phase could be an acceptable model. The surge capacitance should be calculated from the white-box model of the transformer considering the effect of the bushings. To represent the effect of the transformer on the subsequent overvoltages, which are often in the 100 kHz range, a high-frequency model must be applied.

7.3   MATLAB®/SIMULINK® EXAMPLE

In this part, we are exploring the detailed model of a transformer suggested in Figure 7.6. It is worth noting, although the software of the choice here is PSCAD, modeling the case suggested in Figure 7.6 considering mutual inductances is not quite feasible. On the other hand, Simulink provides a user-friendly situation in modeling mutual inductances. Figure 7.9 shows the simulated circuit. Supply, as an input, is a DC voltage source with a resistance of 0.001 ohms. The circuit breaker closes at $t = 0.1$ s. Capacitors C and K, shown in Figure 7.6, are 5.96 and 1.02 nF, respectively. The resistance for R is 8 kΩ. A seven-section detailed model has been studied here. It is possible to form the inductance matrix as shown below. All the elements in the matrix are in mH.

$$
\begin{array}{c}
L_1 \\ L_2 \\ L_3 \\ L_4 \\ L_5 \\ L_6 \\ L_7
\end{array}
\begin{bmatrix}
L_1 & L_2 & L_3 & L_4 & L_5 & L_6 & L_7 \\
1.80 & 1.50 & 1.30 & 0.30 & 0.15 & 0.13 & 0.11 \\
1.50 & 1.80 & 1.50 & 1.30 & 0.30 & 0.15 & 0.13 \\
1.30 & 1.50 & 1.80 & 1.50 & 1.30 & 0.30 & 0.15 \\
0.30 & 1.30 & 1.50 & 1.80 & 1.50 & 1.30 & 0.30 \\
0.15 & 0.30 & 1.30 & 1.50 & 1.80 & 1.50 & 1.30 \\
0.13 & 0.15 & 0.30 & 1.30 & 1.50 & 1.80 & 1.50 \\
0.11 & 0.13 & 0.15 & 0.30 & 1.30 & 1.50 & 1.80
\end{bmatrix}
$$

Figure 7.10 Voltages of seven nodes at $t = 0.15$ s, for the following cases:

Case 1 – Considering mutual inductances from the above matrix

Case 2 – Considering the mutual inductances 10 times more than what is suggested by the matrix

Case 3 – Without mutual induction (changing the matrix to a diagonal form, every element off-diagonal is zero)

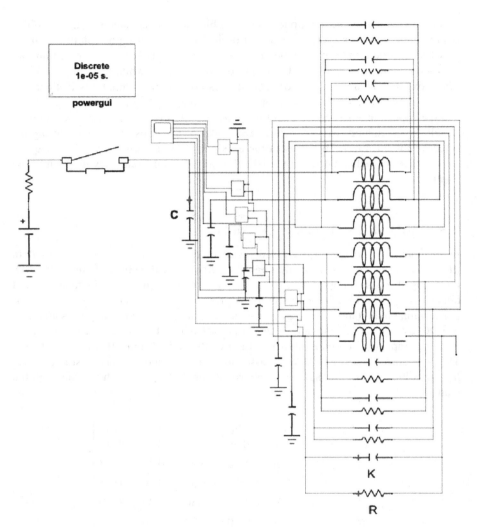

FIGURE 7.9   Simulation system in MATLAB/Simulink.

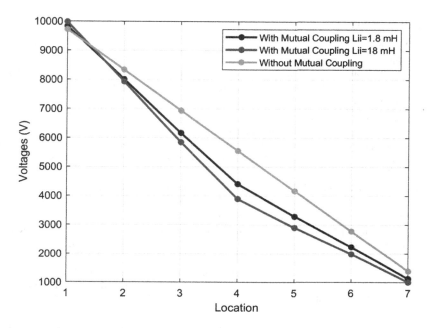

FIGURE 7.10   Simulation results.

Problems

7.1 When there is traveling wave penetration into a winding, what are the nodes along the winding that are more susceptible to experiencing high overvoltages or tensions? Elaborate on your response.

7.2 In the design of a winding, the first 20% of its length has been constructed in a way that its longitudinal capacitance is increased compared to the other 80% length. Ignoring mutual inductance and damping effects, show the following:
   a. Initial distribution of the step wave voltage along this winding $u_0(x)$
   b. Final distribution along the winding $u_e(x)$
   c. Oscillations envelope from the first wave incidence until damping

7.3 In the extended model of a winding, if U is the vector of nodes voltage and $I = [0 \ldots 0, In]$ is the injecting current vector, in the Laplace domain, we will have:

$$(s) = [s^2C + sG + H]$$

   a. Simplify this equation for higher frequencies and draw the resultant electric circuit.
   b. What is the relationship between the distribution of impulse voltage and nodes voltages in this model in higher frequencies?

BIBLIOGRAPHY

1. Juan A. Martinez-Velasco, *Transient Analysis of Power Systems: A Practical Approach*, John Wiley & Sons, 2019.
2. Neville Watson and Jos Arrillaga, *Power Systems Electromagnetic Transients Simulation*, London, United Kingdom: The Institution of Engineering and Technology, 2007.
3. J. C. Das, *Transients in Electrical Systems: Analysis, Recognition, and Mitigation*, The United States of America: McGraw Hill, 2010.
4. J. Nosratian Ahoor, S. Seyedtabai, and G. B. Gharehpetian, "Application of Large Scale Developed Detailed Model for Transformer Transient Studies", *Journal of Iranian Association of Electrical and Electronics Engineers (IAEEE)*, Vol. 14, No. 1, Spring 2017, pp. 83–91.
5. S. M. H. Hosseini, M. Vakilian, and G. B. Gharehpetian "Comparison of Transformer Detailed Models for Fast and Very Fast Transient Studies", *IEEE Transactions on Power Delivery*, Vol. 23, No. 2, April 2008, pp. 733–741.
6. G. B. Gharehpetian, H. Mohseni, and K. Moeller, "Hybrid Modeling of Inhomogeneous Transformer Windings for Very Fast Transient Overvoltage Studies", *IEEE Transactions on Power Delivery*, Vol. 13, No. 1, January 1998, pp. 157–163.
7. M. M. Shabestary, A. J. Ghanizadeh, G. B. Gharehpetian, and M. Agha-Mirsalim, "Ladder Network Parameters Determination Considering Non-dominant Resonances of Transformer Winding", *IEEE Transactions on Power Delivery*, Vol. 29, No.1, February 2014, pp. 108–117.
8. J. Nosratian Ahour, S. Seyedtabaie, and G. B. Gharehpetian, "Modified Transformer Winding Ladder Network Model to Assess Non-Dominant Frequencies", *IET Electric Power Applications*, Vol. 11, No. 4, April 2017, pp. 578–585.
9. Mohammad Amin Sobouti, Davood Azizian, Mehdi Bigdeli, and G. B. Gharehpetian, "Electromagnetic Transients Modelling of Split-Winding Traction Transformer for Frequency Response Analysis", *IET Science, Measurement & Technology*, Vol. 13, No. 9, November 2019, pp. 1362–1371.
10. E. Rahimpour, J. Christian, K. Feser and H. Mohseni, "Transfer Function Method to Diagnose Axial Displacement and Radial Deformation of Transformer Windings", *IEEE Transactions on Power Delivery*, Vol.18, No.2, April 2003, pp. 493–505.
11. CIGRE TB 577A, "Electrical Transient Interaction between Transformers and the Power System. Part 1 – Expertise", JWG A2/C4.39, 2014.
12. CIGRE TB 577B, "Electrical Transient Interaction between Transformers and the Power System. Part 2 – Case Studies", JWG A2/C4.39, 2014.
13. V. Rashtchi, E. Rahimpour, and E. M. Rezapour, "Using a Genetic Algorithm for Parameter Identification of Transformer R-L-C-M Model", *Electrical Engineering*, Vol. 88, No.5, 2006, pp. 417–422.
14. B. Gustavsen and Á. Portillo, "A Damping Factor-Based White-Box Transformer Model for Network Studies," in IEEE Transactions on Power Delivery, vol. 33, no. 6, pp. 2956-2964, Dec. 2018, doi:10.1109/TPWRD.2018.2847725.
15. J. A. Martinez, R. Walling, B. A. Mork, J. Martin-Arnedo, and D. Durbark, "Parameter Determination for Modelling System Transients – Part III: Transformers". *IEEE Transactions Power Delivery*, Vol. 20, No. 3, July 2005, pp. 2051–2062.

8 Ferroresonance

In circuit theory, resonance is defined as a condition where the circuit capacitive impedance is the same as its inductive impedance, that is, the impedances cancel each other at a certain frequency known as resonance frequency. This can happen in circuits with linear elements and the circuits with inputs of alternating signals. In the case of having a nonlinear inductance, the condition will change, and the phenomena is called ferroresonance instead of resonance. The ferroresonance is typically initiated by saturable magnetizing inductance of a transformer and a cable or a transmission line with intrinsic capacitors that are connected to the transformer. In most practical situations, ferroresonance results in dominated currents, but in some operating modes, it may cause significantly high values of distorted winding voltage waveform, which is typically referred to as ferroresonance. Although the occurrence of a resonance involves a capacitance and an inductance with a known resonance frequency, there is no definite resonance frequency in this case. In this phenomenon, more than one response is possible for the same set of parameters, and drifts or transients may cause the response to jump from one steady-state response to another. Its occurrence is more likely to happen in the absence of adequate damping. Considering the classifications presented in Figure 1.3, the ferroresonance can be considered in the group of sustained and temporary overvoltages. Temporary overvoltages are undamped or little damped overvoltages in power frequency. Their duration is in the range of seconds or minutes, and therefore, cannot be considered as electromagnetic transients, which have been studied in previous chapters. They can typically happen due to ground faults, load rejection, energization of unloaded transformers, or a combination of the mentioned cases. We aim in this chapter to study this important phenomenon in power systems. Research on ferroresonance in transformers has been conducted over the last 80 years. The word "ferroresonance" first appeared in the literature in 1920, although studies on resonance in transformers were published as early as 1907. Practical interests had been shown in the 1930s when it was shown that the use of series capacitors for voltage regulation could cause ferroresonance in distribution systems. In normal conditions, the three-phase windings of power system transformers are in parallel with network capacitors. During transients due to switching, the windings can get in series with capacitors. Also, the high voltages appeared due to transient conditions that may saturate the transformer coil that has got in series with the system capacitor. There are cases where the system voltage transformer (VT) and intrinsic capacitors can form a resonance circuit. In this case, the saturable winding of the VT and the capacitors are in parallel. The major concern in cases of series or parallel ferroresonance is abnormal high voltages that appear in the system due to the formation of the resonance circuit. Parameters such as length of line and loading of a transformer are considerable in the appearance of ferroresonance.

DOI: 10.1201/9781003255130-8

8.1   LINEAR CIRCUIT RESONANCE

In order to analyze this condition, we start with the following linear circuit as shown in Figure 8.1.

For this circuit, we can write the following equations based on KVL.

$$u_0 = \frac{1}{jC\omega}I + j\omega LI \tag{8.1}$$

$$u_0 = -j\frac{1}{C\omega}I + j\omega LI \tag{8.2}$$

$$u_0 + j\frac{1}{C\omega}I = j\omega LI \tag{8.3}$$

In this circuit, the phasor of the voltage u_0 is aligned with $j\frac{1}{C\omega}I$ and $j\omega LI$. Therefore, we can write the following equation for the magnitudes.

$$u_0 + \frac{1}{C\omega}I = \omega LI \tag{8.4}$$

Now, we can plot Equation 8.4 as shown in Figure 8.2. As it can be seen in this figure, the two intersected lines are presenting ωLI, and $u_0 + \frac{1}{C\omega}I$ (red lines). There is the second set of lines, which are the parallel lines that are presenting the resonance situation at $\omega = \frac{1}{\sqrt{LC}}$ (blue line). Finally, we have another set of lines that are perpendicular (orange line and green line). They represent the condition where $\omega \to 0$ and and $\omega \to \infty$, respectively.

The right-hand side of Equation 8.4 is a line with the slope of ωL as well as the left-hand side, which is a line with the offset of u_0 and the slope of $\frac{1}{C\omega}$. It is obvious that first we consider ω constant and I as a variable. The intersection of these two lines (red lines) is the operating point of the resonance circuit as shown in Figure 8.2. Let us take a look at the circuit operating point as ω changes.

1. First, let us consider the case $\omega \to \infty$. Here, the slope of the line ωL is an infinite number, and the line $u_0 + \frac{1}{C\omega}I$, will be changed to u_0. The vertical

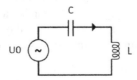

FIGURE 8.1   Linear resonance circuit.

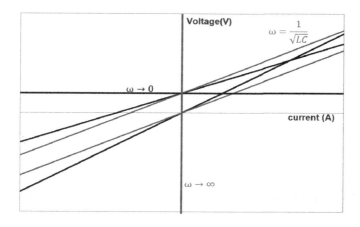

FIGURE 8.2   Volt-amp characteristic of a linear resonance circuit.

green line shown in Figure 8.2 demonstrates this condition. In this situation, the circuit current is zero.

2. Reducing ω results in the reduction of the slope in line $\omega L I$ and increasing the slope of $u_0 + \dfrac{1}{C\omega} I$ line. This condition is illustrated as two intersected red lines in Figure 8.2. There will be an electric current flowing through the circuit in this condition as shown with the intersection of the lines.

3. More reduction in ω leads to a special case known as resonance, that is, $= \dfrac{1}{\sqrt{LC}}$. This situation will lead to a resonance situation illustrated with two parallel lines (blue lines in Figure 8.2). In this case, due to the cancelation of impedances in the circuit, the current is infinite.

4. More reduction in frequency ($\omega < \dfrac{1}{\sqrt{LC}}$) results in the intersection of these lines in the third quadrant of the plane shown in Figure 8.2. This means that a limited amount of current will flow in an opposite direction through the circuit.

5. In the case of $\omega \to 0$, the intersection of these two lines is at (0,0). In this situation, the current is zero.

As it can be seen, u_0 is not affecting the resonance in the circuit. Figure 8.3 summarizes the above discussion and illustrates how ω affects the current in the circuit.

Let us consider the situation where $\omega \to \infty$ and the current is zero. As ω decreases, the current will increase. In case of $\dfrac{1}{C\omega} = L\omega$, current will be $\to \infty$. After that, by decreasing ω, the current starts flowing in the opposite direction. Current reduces as ω decreases as shown in Figure 8.3 on the left. Figure 8.3 on the right shows the current amplitude (orange graph) and the effect of adding a resistance in the circuit (blue graph). Adding a resistance will limit the current due to resonance.

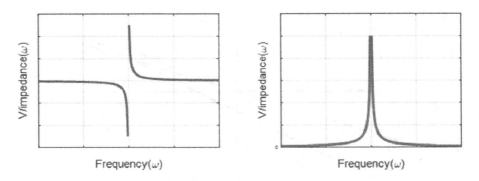

FIGURE 8.3 Current frequency characteristic of a resonance circuit (current on left, magnitude of current on right).

8.2 NONLINEAR CIRCUIT RESONANCE

Figure 8.4 shows a series resonance circuit with a nonlinear inductor.

Figure 8.5 shows the relationship between the voltage across the nonlinear inductor and current flowing through the resonance circuit with ω as a changing parameter. It can be seen that at a constant current, an increase in ω results in an increase in the voltage across the nonlinear inductor. In this figure, yellow graph is related to highest ω and blue graph is related to lowest ω.

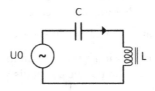

FIGURE 8.4 Nonlinear series resonance circuit.

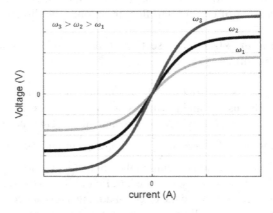

FIGURE 8.5 Nonlinear inductor volt-amp characteristic.

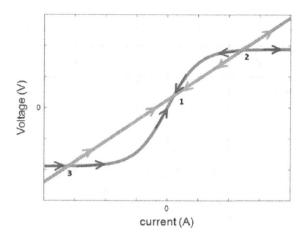

FIGURE 8.6 Volt-amp characteristic of the nonlinear circuit.

In the circuit shown in Figure 8.4, there is no resistance and the phasor of the voltage u_0 is again aligned with $j\dfrac{1}{C\omega}I$ and $f(I)$. Therefore, considering only the magnitudes, we can write the following KVL.

$$u_0 + \frac{1}{C\omega}I = f(I) \tag{8.5}$$

where $f(I)$ is voltage across the nonlinear inductor. Figure 8.6 shows the line $u_0 + \dfrac{1}{C\omega}I$ and the nonlinear characteristic $f(I)$ in the u-I plane. Comparing this figure with Figure 8.2, it can be seen that in this case, instead of one operating point, there will be three possible operating points. Points 1 and 2 are in the first quadrant of $u - I$ plane, and point 3 is in the third quadrant of this plane. In the following subsection, it will be shown that only one of these operating points is a stable operating point.

8.2.1 STABILITY ASSESSMENT OF OPERATING POINTS IN A NONLINEAR RESONANCE CIRCUIT

In this subsection, we are going to consider the perturbation method to study the stability of the operation of the circuit shown in Figure 8.4. As can be seen in Figure 8.5, we have three operating points (i.e., the intersections of the nonlinear characteristic with the linear characteristic). At each operating point, we will consider a positive and negative perturbation and will discuss the behavior of the operating due to these perturbations.

a. Operating point 1

Assume that there is a positive perturbation of ΔI in current at this operating point. As it can be seen in Figure 8.6, voltage across the inductor (i.e., (I) is greater than $u_0 + \dfrac{1}{C\omega} I$, which is the linear characteristic (i.e., $u_0 + \dfrac{1}{C\omega} I < f(I)$. Therefore, the current would flow from the inductor to the rest of the circuit, and as a result, the current will decrease till it reaches point 1 again.

Now, we assume a decrease of ΔI in the current. In this case, as can be seen in Figure 8.6, we have:

$$u_0 + \frac{1}{C\omega} I > f(I)$$

Then, the current will flow from the part of the circuit with the line characteristic to the nonlinear part. Therefore, the current will increase again, and we will reach point 1 again. As a result, it can be said that operating point 1 is a stable point from both sides.

b. Operating point 2

Assume that we have a positive perturbation of ΔI in current at this operating point. In this case, as can be seen in Figure 8.6, the voltage across the nonlinear part of the circuit, that is, $f(I)$ would be less than the linear characteristic, that is:

$$u_0 + \frac{1}{C\omega} I > f(I)$$

Therefore, the current would flow more from the linear side to the inductor side; as a result, the current will increase more and more and there is no stable operating point for this condition.

Now, we assume a decrease of ΔI in current. In this case, as can be seen in Figure 8.6, we have:

$$u_0 + \frac{1}{C\omega} I < f(I)$$

In this case, the current will flow from the nonlinear part of the circuit to the linear part of the circuit, which results in current reduction. Therefore, the operating point of the system reaches point 1 and not point 2. As a result, it can be said that operating point 2 is an unstable point.

c. Operating point 3

Again, assume that there is a positive perturbation of ΔI in current at this operating point. As can be seen in Figure 8.6, the voltage across the inductor, that is, $f(I)$ would be less than the linear characteristic, that is:

$$u_0 + \frac{1}{C\omega} I > f(I)$$

As a result, the current would flow from the linear part of the circuit to the inductor, and therefore, the current will increase till it reaches operating point 1 instead of point 3.

Now, consider a negative perturbation of ΔI in current at this operating point. In this case, as can be seen in Figure 8.6, we have:

$$u_0 + \frac{1}{C\omega} I < f(I)$$

The current will flow from the nonlinear part of the circuit to the linear part of the circuit, which results in a decrease in current. Therefore, the system operating point reaches point 3 once again. Consequently, operating point 3 is a stable point for a negative disturbance but an unstable point for a positive disturbance.

In the end, it can be said that there are three possible operating points of the circuit for a constant ω and a constant u_0.

- Point 1: Non-ferroresonance stable operating point in which the circuit is working in an inductive mode, with a lagging current and low voltages.
- Point 2: an unstable operating point.
- Point 3: Ferroresonance stable operating point in which the circuit is working in a capacitive mode, with leading current and high voltages.

The effect of ω and u_0 changes.

We saw previously that u_0 does not affect the appearance of a resonance in the linear circuit shown in Figure 8.1. In the case of a nonlinear circuit, we will see that both ω and u_0, will affect the system resonance and operating points.

8.2.1.1 Effect of ω

Figure 8.7 shows the nonlinear characteristic of the inductor alongside the line characteristic presenting the source and capacitor. For the sake of simplicity assume, the nonlinear inductor is frequency independent, that is, $f(I)$ and will not change with ω. The goal is to study the effect of changes in ω on the operating point of the circuit. In general, the figure shows conditions where there is no meaningful operating point (i.e., no intersection), which is illustrated in green. The tangential red line will

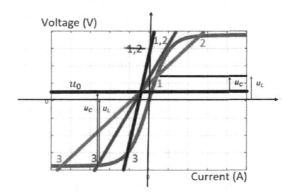

FIGURE 8.7 Effect of ω on the operating point of a ferroresonnance circuit.

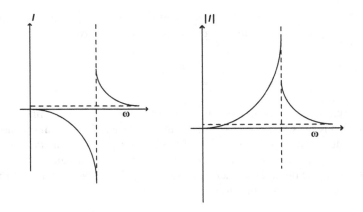

FIGURE 8.8   Effect of ω on the ferroresonance circuit current.

provide two operating points with the same coordinates. The brown line is a condition where there are three different operating points.

First, we consider the case $\omega \to \infty$. In this case, the line equation $u_0 + \dfrac{1}{C\omega} I$ will tend to u_0, and as it can be seen in Figure 8.7, there will be only one operating point, that is, point 1 (black line). As ω decreases, we will have three operating points' ①, ②, and ③ (orange line). This case is similar to the one presented in Figure 8.6. The important result of this case is that the voltages across the inductor and capacitor are not high and $u_C < u_L$, which is shown in this figure. More reduction in ω results in the merging of operating points 1 and 2 (the tangential red line). In this condition, we have two operating points. One is point 3 and the other one is Point (1,2). The next step is so critical. A slight reduction in ω means that the line $u_0 + \dfrac{1}{C\omega} I$ will have only one intersection with $f(I)$, and as a result; we have only operating point 3. This happens with a sudden jump from the old operating point (1,2) to the new operating point. The important result of this case is that the voltages across the inductor and capacitor are very high, that is, overvoltages, and $u_C > u_L$, which is shown in this figure as well. These overvoltages due to ferroresonance can damage devices, especially the capacitors.

Figure 8.8 summarizes the above discussion and illustrates how ω affects current in a ferroresonance circuit, which can be compared with Figure 8.3.

8.2.1.2   Effect of u_0

Figure 8.9 indicates the nonlinear characteristic of the inductor, that is, $f(I)$ alongside the line $u_0 + \dfrac{1}{C\omega} I$. For the sake of simplicity, assume that $f(I)$ will not change due to changes in u_0, and we want to study the effect of changes in u_0 on the operating point of the circuit.

First, we consider the case, which is similar to the one presented in Figure 8.6. In this case, we have three operating points ①, ②, and ③, as shown in Figure 8.9. By increasing u_0, the operating points ① and ② will merge into one point named (1,2).

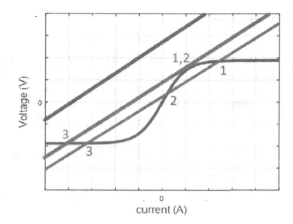

FIGURE 8.9 Effect of u_0 changes on the operating point of a ferroresonnance circuit.

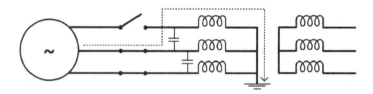

FIGURE 8.10 Three-phase power supply with one-phase contact trip due to a breaker malfunction.

In this case, we have two operating points, that is, 3 and (1,2). A slight change of u_0 leads to a jump from operating point (1,2) to operating point ③. Again, the important conclusion of this case is that the voltages across the inductor and capacitor are very high, that is, overvoltages, and $u_C > u_L$. These ferroresonance overvoltages can damage devices, especially the capacitors.

8.3 FERRORESONANCE PRACTICAL CASES IN POWER SYSTEMS

8.3.1 ONE PHASE OPEN

Figure 8.10 shows a three-phase power system. There is a single-phase trip on the circuit breaker due to a malfunction. Consider the intrinsic capacitor between transmission lines (shown in the figure). A single-phase LC circuit will be formed due to the asymmetrical breaker operation. This will be similar to the condition discussed in Figure 8.4. The power delivered to the load will reduce due to the breakage of one of the phases, and as a result, the frequency of the source can increase similar to the case discussed in Section 8.2.2.1. Therefore, the combination of the capacitor and the nonlinear inductor in series will form a ferroresonance circuit, which can lead to overvoltages. The overvoltage in capacitor ($u_C > u_L$) means an overvoltage between two transmission lines, which can result in a breakdown and flashover between the lines.

8.3.2 Overload Relay Malfunction

Overcurrent relays must disconnect the load in order to balance the load and supply. In case of occurrence of malfunction in these relays, the load will consume excessive power resulting in frequency reduction in the whole system. At the same time, in steam power plants, the steam valve may reduce the input power leading to more reduction in frequency. These conditions are the same as the condition discussed in Section 8.2.2.1, and as a result, a ferroresonance can occur in system LC combinations.

8.3.3 De-energization of One Circuit of Two-circuit Transmission Line

In high voltage transmission lines, there are potential transformers (PTs) directly connected to the line. In two-circuit transmission lines, in case of having one circuit de-energized, as shown in Figure 8.11, the PT will be connected to the line. On the other hand, we have the capacitances between energized and de-energized lines, as shown in the figure. Therefore, the combination of the voltage of the energized line, capacitances, and the inductance of the PT can form the circuit shown in Figure 8.4. As a result, a ferroresonance can occur in this combination.

8.3.4 Single-Phase Fault and Capacitive Voltage Transformer

Figure 8.12a shows the connection of a capacitive voltage transformer (CVT) to the line. The equivalent circuit of this connection can be modeled by the circuit shown in Figure 8.12b. The Thevenin voltage has been represented by U_0. Now, assume

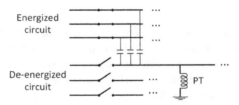

FIGURE 8.11   One circuit of a two-circuit transmission line is de-energized.

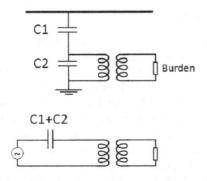

FIGURE 8.12   (a) Single-phase fault and CVT and (b) the equivalent circuit.

that there is a single-phase fault; therefore, we expect to have a voltage rise of $\sqrt{3}$ pu in healthy lines. This voltage rise/change in U_0 is similar to the case discussed in Section 8.2.2.2. As a result, a ferroresonance can occur in this combination, as well.

8.4 SIMULATIONS

The following circuit is being modeled in MATLAB/Simulink to study the ferroresonance phenomenon.

The system parameters are (Figure 8.13):

Winding 1: $V_1 = 5.773$ kV (rms-LG), $R = 0.667$ MΩ, $L = 2998$ H
Winding 2: $V_1 = 110$ V (rms-LG), $R = 242$ Ω, $L = 0.771$ H

Saturation curve data ([i1(A), phi1(Wb); ...]): [0 0 ; 0.0408 0.1681 ; 0.0735 0.3542 ; 0.1225 0.5346 ; 0.2123 0.7076 ; 0.6287 0.8653 ; 2.1066 1.0383 ; 4.1805 1.2114 ; 6.4912 1.3575 ; 10.7778 1.5344]

A single-phase fault is applied to the circuit. Fault happens at $t = 0.1$ s, and at $t = 0.2$ s, the fault gets cleared. As soon as the fault is applied to the system, the phase A of the breaker trips, and at $t = 0.2$ s, it gets connected again. Therefore, voltage waveforms shown in Figure 8.14 are presenting the ferroresonance phenomenon.

Problems

8.1 In the single-line diagram shown below, overload has not been detected by the overcurrent relay, and therefore, the generator has not been tripped. The turbine closes the steam valves simultaneously. After this incident, the CVT is damaged. Explain the reason behind the CVT damage.

8.2 The line shown in the diagram is compensated by a series capacitor. If lightning hits the line, explain the conditions that can lead to ferroresonance in the system.

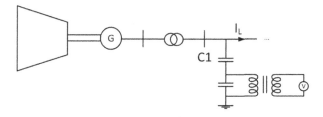

FIGURE P8.1   CVT damage case single-line diagram.

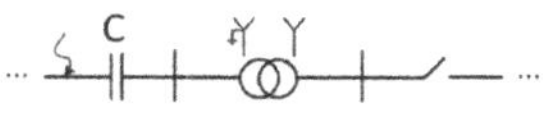

FIGURE P8.2   Ferroresonance case.

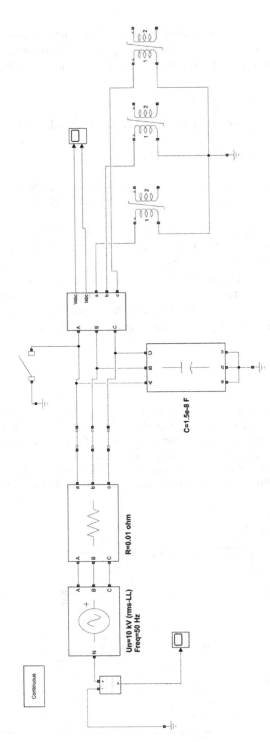

FIGURE 8.13 Simulated power system for ferroresonance studies.

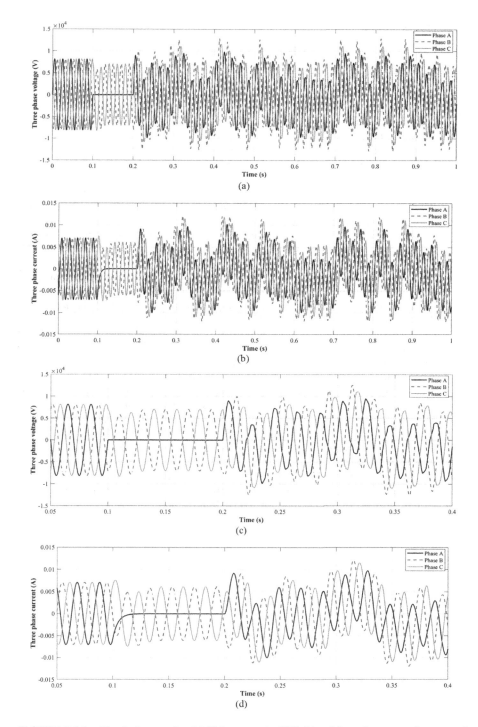

FIGURE 8.14 Simulation results. (a) Voltages at the HV side of the voltage transformer and (b) currents.

BIBLIOGRAPHY

1. J. R. Marti and A. C. Soudack, "Ferroresonance in Power Systems: Fundamental Solutions", *IEE Proceedings C Generation Transmission and Distribution*, Vol. 138, 1991, pp. 321–329.
2. Afshin Rezaei-Zare, M. Sanaye-Pasand, H. Mohseni, S. Farhangi, and R. Iravani, "Analysis of Ferroresonance Modes in Power Transformers Using Preisach-Type Hysteretic Magnetizing Inductance", *IEEE Transactions on Power Delivery*, Vol. 22, No. 2, April 2007, 919–929.
3. Radwan Taha AL-Bouthigy, Ahmed Sayed, and Mazen Abdel-Salam, "Ferroresonance Phenomenon in Electrical Power Transformers", 2012, LAP LAMBERT Academic Publishing.
4. R. Minkner, J. Schmid, H. Däumling, et al., *Ferroresonance Oscillations in Substations: With Inductive Voltage Transformers in Medium and High Voltage Systems*, VDE Verlag GmbH.
5. H. Radmanesh and G. B. Gharehpetian, "Ferroresonance Suppression in Power Transformers using Chaos Theory", *Electrical Power and Energy Systems*, Vol. 45, 2013, pp. 1–9.
6. Radmanesh, Hamid and Fathi Seyed Hamid, "Analyzing Ferroresonance Phenomena in Power Transformers Including Zinc Oxide Arrester and Neutral Resistance Effect", *Applied Computational Intelligence and Soft Computing*, Vol. 2012, January 2012, Article No.: 16pp 16 https://doi.org/10.1155/2012/525494.
7. Hamid Radmanesh and G. B. Gharehpetian, "Ferroresonance Suppression in Power Transformers using Chaos Theory", *International Journal of Electrical Power and Energy Systems*, Vol. 45, No. 1, February 2013, pp. 1–9.
8. A. Abbasi, S. H. Fathi, G. B. Gharehpatian, A. Gholami, and H. R. Abbasi, "Voltage Transformer Ferroresonance Analysis using Multiple Scales Method and Chaos Theory", *Complexity*, Vol. 18, No. 6, July/August 2013, pp. 34–45.
9. Mehran Esmaeili, Mehrdad Rostami, G. B. Gharehpetian, and Colin P. McInnis, "Ferroresonance After Islanding of Synchronous Machine-Based Distributed Generation", *Canadian Journal of Electrical and Computer Engineering*, Vol. 38, No. 2, 2015, pp. 154–161.
10. Nattapan Thanomsat, Boonyang Plangklang, and Hideaki Ohgak, "Analysis of Ferroresonance Phenomenon in 22 kV Distribution System with a Photovoltaic Source by PSCAD/EMTDC", *Energies*, Vol. 11, 2018, 1742. doi:10.3390/en1107174

Index